A Blessed Life

One World War II Seabee's Story

Tamra McAnally Bolton 2/6/20

Tamra McAnally Bolton

R. Chance

*"Those affected by the uncertainty of war will be moved by this book...*A Blessed Life *is both a touching personal recollection of an East Texas boy forced to become a man in the heat of war as well as an illustration of the journeys of millions of Americans during World War II, afraid to leave home, but resolved to defend the values America holds dear."*

Lieutenant Colonel Benjamin Bryant,
US Air Force, War Veteran,
and fellow East Texan

Tamra McAnally Bolton uncovers the long-buried war-time memories of her 96-year-old father over coffee and raisin pie at the kitchen table of the long-time family home near Jacksonville, Texas. This surprisingly gentle story of an East Texas farm boy going to fight for his country wraps around the reader like a homemade quilt. You're not going to want to put this book down, so make your pot of coffee ahead of time.

Brad Maule,
actor and singer best known for his role as
General Hospital's Dr. Tony Jones,
Acting/Directing and Filmmaking professor
Stephen F. Austin State University in
Nacogdoches, Texas.

A must-read account of captivating conversations between a daughter and her veteran WW II father. The unfolding of history and the cast of characters he met in service to his country is undoubtedly one of the most descriptive accounts of the action, sacrifice, and service of these treasured men ever written.

Lieutenant Colonel Mark Green,
US Army, retired,
author of numerous books including
Step Out, Step Up: Lessons Learned from a Lifetime of Transitions and
Military Service; Warrior's Code 001: 7 Vital Steps to Resiliency and
I Am Still Here, an international best-seller.

A Blessed Life *is a precise and accurate reflection of being a WWII soldier, as told by the soldier himself. Through his daughter's many persistent and sequential interviews with her father, the reader experiences this Navy Seabee's mindful images, thoughts and emotions as he leaves his family to enter WWII military service; as he and his fellow soldiers live out their military responsibilities on Iwo Jima i.e. "We build, we fight" and as he returns home after the war. This book well portrays the many mental and physical challenges that the Seabee, Navy, and Marine forces experienced fighting the Japanese in the South Pacific Theater during WWII and vividly describes Japan and the conditions there after the war.*

Dr. James F. Harbaugh, Sergeant Major,
U.S. Army Reserve, Retired

Also By Tamra

When I Was Small

A Blessed Life: One World War II Seabee's Story.

Poverty Ridge Press
1090 CR 3905
Jacksonville, TX 75766
Email: povertyridgepress47@gmail.com

Published in the United States of America.

Cover photo by: Tamra M. Bolton
Back cover author photo by: Melanie Roos
Shadow image photo by: jorik/Shutterstock
Editors: Sherry L. Bryant and Wanda Baham Sturrock
Interior Layout: Deena Rae Schoenfeldt — E-Book Builders

Print ISBN: 978-1-7343445-4-7
eBook ISBN: 978-1-7343445-5-4

This book is dedicated to the men of the 31st USNCB who served from 1942 to 1946

Acknowledgments

One of the things I have learned while working on this project is that no book is the sole creation of the author. It takes many people and their wisdom and insight to birth a book.

I am deeply grateful to Sherry L. Bryant and Wanda Baham Sturrock whose sharp eyes and skills in editing are only surpassed by their extreme patience and encouragement. Also thanks to Deena Rae at E-BookBuilders who is a whiz at book design and layout and making me look good.

I am thankful to my husband Marc, who gave me the space I needed to work on this project for over six years, and to my family for always being my biggest cheerleaders.

Most of all, I am thankful to God who gave me the ability to do what I do and who blessed me with an amazing earthly Dad who shows me every day what being a hero really means.

AUTHOR'S NOTE

All of the events described in this book are based on the memories of my father, Stuart McAnally, as told to me in thirty-four recorded interviews. The events are told from his point of view, in his own words as he experienced them. He also credits some memories to fellow 31st members and relays them as such.

The dates and events were carefully researched and substantiated by the original 31st Construction Battalion book and other corroborating papers and mementos. Some of these are featured in the photo section of the book. The recorded interviews are part of the permanent archives at The National Museum of the Pacific War in Fredericksburg, Texas. These recordings are accessible to the public by request when you visit the museum.

The portions of the book in italics are based on Dad's childhood memories and his return from World War II. These are intended to give the reader some insight into his background and a glimpse of life during the Great Depression and the 1940s.

A reprint of a Parade.com article, first posted on June 16, 2017, tells the story of what kind of man my father became after the war. You can find this at the back of the book, or you can go online and read the article with the photos included at: **https://parade.com/579500/tamrabolton/celebrating-68-years-of-fatherhood/**

Also located at the back of the book is a glossary that explains certain military terms and perhaps unfamiliar abbreviations or words used in dialogue.

It is my hope that you will enjoy this journey through Dad's experiences and discover how he created a truly blessed life.

CONTENTS

A Blessed Life

One World War II Seabee's Story

The honeysuckle is early. I smell it through the open windows of the traveling bureau, even with seven or eight smelly service men surrounding me. Inside the cavernous vehicle though, we have plenty of room. I was lucky to find enough guys heading my way to split the cost of the ride. Ten dollars apiece to take us all the way from Houston to Tyler. A bus would take too long and I didn't have the extra money for a train ticket, so this noisy contraption would have to do.

The others joke and trade stories, but I just listen, watching the busy Houston streets fade into scattered storefronts, and then open countryside. Weathered farmhouses with gardens and old men plowing glide by as the bureau noses northward. I can hardly believe I will soon be there…it has been so long.

I close my eyes, pretending to doze, and try to picture it… wondering how much has changed in the last three years. Everything changes, even if you're not there to see it. I hope the Palace Theatre is still open. I promised that Martin girl I would take her to the picture show when I got back. I remember thinking "if I come back" when I told her that, but I never said it out loud.

A bump in the road jars everyone, and a couple of the Marines sitting next to me begin to good-naturedly harass the older bearded driver. Everyone is in a jovial mood today. Everyone is going home. Most of the boys are headed to Tyler, but I'm bound for tiny Jacksonville, just south of there. But the closer we get, the more my stomach starts seizing up and the rest of them get steadily quieter, too. Don't know why, maybe all of us are thinking the same thing…about how lucky we are to be almost home when so many didn't make it back.

We are all changed, even if we don't look a whole lot different on the outside. Our insides are different. It's like picking up where we left off, except the rules have changed and the players are different and everyone understands the rules but you.

The first glimpse of town is the old basket factory smokestack belching black soot into the blue April sky. A new service station has sprouted in Old Man Dewey's cornfield. Guess he won't be plowing that field anymore, I can't help thinking as we drive by. I wonder if Mr. Dewey is still around, maybe that explains the station. A lump in my throat begins to grow as I think about what I might face and who might not be around anymore.

Grandma had passed while I was still on Iwo Jima. I cried when I got the letter from Mama. I remember thinking that it was a good thing I waited till everyone went to the chow hall before I opened it.

When the driver pulls up to the corner near the icehouse, I grab my sea bag, mumble a thank you, and wave as the others drive on.

As the traveling bureau disappears over the rise, I heft the canvas bag to my shoulder and look around. Ready to walk the three miles home, I stand there on Commerce Street, looking at the town and I am relieved to see that little has changed. The bank and drugstore look just like I remember, but I can't shake the feeling that I am out of place. Now that I am actually standing here in Jacksonville, I have the strangest feeling that I am going to wake up back in my tent on Iwo with that lump of homesickness in my stomach.

But, it is real. I wrap my fingers around the closest thing I can find…the stop sign pole on the corner. The feel of the cool metal on my palm reassures me that it's real. I am alive, and I am going home.

Chapter One

"Back in a minute!" I hollered to my husband, standing in the doorway of his shop.

He waved without looking up from the tool bag he'd been digging through. He had been out there all morning trying to resurrect our ailing mower, hoping it would make one more season.

I put the bowl of cream peas and a pan of cornbread in a box, loaded it in the Jeep and drove the half mile to Mom and Dad's, expecting Dad to greet me at the door with his usual grin and greeting.

"Well, well," he'd always say, "if it isn't the food fairy with her latest delivery!"

Mom and Dad don't always eat like they used to. Like a lot of older folks, they only eat once or twice a day, sometimes just a snack instead of a regular meal like they need. When Dad is involved with one of his projects, though, he does eat more often.

When I got to the house, I didn't see Dad. Balancing the box in my hands, I knocked on the kitchen door, and opened it a crack.

"Hello!" I called.

No one answered. I set the dishes on the kitchen table, called out again, then peeked into the living room. Mom was sound asleep, lent back in her recliner, her Wheel of Fortune puzzle book on her lap. Dad's chair was empty, which meant he was probably back in the sunroom where he liked to sit during the day.

I walked the length of the house to the sunroom and found him in his chair, softly snoring. The burgundy recliner made his head of white hair look even brighter and I could see from his outfit that he must have been working on some project either outside or up at the barn. I smiled at his favorite green plaid flannel shirt, "air-conditioned" at the elbows and his stained khakis. He liked wearing clothes for comfort and didn't think it necessary to dispose of them if they were "good and broken in" as he liked to say. I stood there for a moment admiring his still handsome, but weathered profile before deciding to wake him.

"Dad?" I said.

His eyes slowly opened in an unfocused stare for a second or two, then that familiar smile spread from his eyes to his mouth when he looked over and noticed me standing in the doorway.

"Hey…when did you get here?" he asked.

"Oh, just a little while ago." I grinned. "Looks like you've been working too hard again."

He sat up in his chair and swiped his hands over his face as if to wipe the last bit of sleepiness away.

"Oh…I've just been up at the barn workin' a little on a project."

He always has projects going, from building bookcases and birdhouses to clearing fencerows and rebuilding fence.

"I brought some fresh cream peas and cornbread for supper."

"Boy, that sounds like a winner!" He grabbed the arms of his chair to get up. "I think I've built up an appetite in this chair."

After struggling a bit, he finally got to his feet and we headed back through the house.

"Well, there's plenty," I said, "so I hope you're hungry, 'cause we both know Mama won't eat much."

"Nah…she's a carnivore; if it isn't meat, she's not too interested."

"Not any food better than peas and cornbread," I thought out loud.

"Mmmm! Mmmm! That's what I say!" he agreed. "I'll fix me a plate right now, while it's hot."

Dad shuffled to the knotty pine cabinet and pulled out a white and gold Corning ware plate, then got a knife and fork from the drawer beside the old white farmhouse sink.

"Would you like a glass of water?" I asked.

"I guess so."

Dad forgets to drink water. We tease him that he must be half camel because he can go so long without drinking water. Mostly, he just drinks coffee. Sometimes during a heat wave he drinks some cold water, but that is the exception. We've been trying to get him to drink more water since the doctor told him being dehydrated was one of the reasons for the dizzy spells he's been having lately.

I fixed him a glass of water and set it on the wooden dining table, then watched as he heaped the peas and cornbread onto his plate. He sliced a ripe red tomato from his garden and peeled and quartered a green onion he said he had pulled just yesterday.

When he sat down, Dad rapped his knuckles on the table - a gesture he has used since our childhood to get our attention before he says the blessing. I hurried to pour my cup of coffee and then took my place at the table. He bowed his head.

I never tire of hearing Dad pray. His praying is always full of thankfulness and his own turns of phrase so characteristic of him and no one else. Sometimes I find myself wondering what God thinks of Dad's prayers. My guess is…God thinks they are pretty special too.

• •

On days I can stay for a while, I linger at the table while Dad is eating, keeping him company. Oftentimes, he will tell stories from "the old days" as he calls them, boyhood adventures and family tales, ones that never fail to amuse me.

Earlier, when I came into the sunroom, I'd noticed he had his old World War II 31st CB book open by his chair.

"Dad…I saw that your CB book was out. Looking somebody up?"

He shook his head.

"Nah, just doing a little wool-gatherin', as Mama used to say. When I doze off during the day I sometimes dream a bit, but I don't have many dreams about that time any more. Sometimes though, I dream about times 'before the war.'"

He always divides his life, referring to time as before the war and after the war. When I listen to other veterans talk about their lives, they often use the same terms.

"Is that what you were dreaming about today?" I asked, as I handed him the pepper shaker. "Or was it one of your 'working dreams'?"

Dad frequently has the craziest dreams about working as a house painter, which was his occupation for over fifty years. Nutty stuff, like driving a house down the road and steering it by the louvers, or putting up the fanciest wallpaper only to have it peel right off again, or painting a house and the paint bubbling off the house while he stands looking at it, the brush still in his hand. I usually ask about them because they are always entertaining.

He took a bite of cornbread and chewed for a minute while he looked out the big picture window by the table.

"You know," he said, "when you caught me in my chair I was dreamin' about coming home from the war."

"Really?" My ears perked up.

"You know, I've had that same dream several times, but I never make it all the way home before I wake up." He looked wistful as he stared out the window. "I would like to see Mama…but I always wake up before I get there…"

I knew he was lonesome for her, even though she'd been gone for almost thirty-five years. He was closer to her than any of the other five boys had been. Dad always stopped by to visit with her every day after work for as long as she lived.

He looked back at his plate and put his fork down. I could feel a story coming.

"You know…that dream got me to thinking about when we found out we were in the war. Dad, Floyd and myself were down in the Big Thicket, huntin' and trappin'. We had been down there for a couple of months and had planned to start heading back along about the middle of December so we'd be home in time for Christmas.

"We loaded the wagon and lit out for home along about the 17th. Took us several days to get home and when we got there on the 21st, I think it was, that's when we heard the news.

"December 7th changed the world. We were about to be in the thick of things. What a shock to come home to. Instead of lookin' forward to celebrating Christmas, you were lookin' at going to war. Yeah…those were hard times, but now, I realize the country was workin' as one and everyone tried to pull their own weight…at least most did."

We talked for a while about how much things had changed since then, and the current political situation. An hour slipped by before I knew it. I picked up Dad's empty plate and took it to the sink and then began making Mama a plate for later.

"What are you gonna do the rest of the day?" I asked as I finished fixing Mama's plate and put it in the fridge.

"Well…think I'll just take it easy. I had a pretty big mornin'." He said as he stood and stretched.

"Sounds like a good idea to me. I'm going to straighten the kitchen and put the dishes in the dishwasher before I leave. Hopefully, I'll see you later for coffee," I said.

"Lord willin'," Dad answered as he headed back towards the sunroom.

I smiled.

When I finished cleaning up, I picked up my keys and locked the door on the way out.

• •

We did get to enjoy our coffee time that afternoon and many more after that, but the spring and summer can be busy times around the farm, with the calving, hay-making and such. With all the activity it was a good while before we slowed down long enough to talk some more about the war and Dad's memories.

• •

One particularly fine fall morning, some months later, I called him to ask if we could make some more recordings of his war time experiences.

"Of course…and we can drink some coffee too," he said.

I gathered my camera and my notebook, and met him a little while later on the back patio. Dad had already made the coffee, so after we fixed ourselves a cup, we just sat for a while admiring the view.

The green pasture was the perfect backdrop for the bits of color that were starting to paint the woods along the creek and the trail up to the ponds. A light north wind brought with it the fragrant pine scent and smell of the oak leaves that were already starting to cover the ground around the barn. Looking at the weathered structure that held so many fond memories, it was hard to believe that Dad built it way back in 1949. Even today, it is still as solid as ever - as much a part of our landscape as the trees and hills. A crow caw broke the morning quiet as we sipped our coffee in comfortable silence. When he was ready, he began to talk…

"When I was in the ninth grade, one of my teachers had just come back from traveling all over Europe and Japan. He said 'there were lots of things in Japan they wouldn't let tourists see.' He said he 'knew war was coming – the sabers were rattling in both Europe and Japan.' He told us, 'You boys will be just the right age to catch it.' Boy…was he right.

"That was in 1938. In August 1941, I turned eighteen, and then Pearl Harbor happened in December.

"I signed up like all young men did then. The Marines wanted me, but I wanted to join the Seabees. One big officer at the enlistment place tried his best to talk me into joining, but I insisted on being a Seabee. I believe that if I had joined the Marines, I would have probably never come home."

While Dad's dark brown eyes scanned his beloved farm, a shadow washed over his face that was hard to decipher, maybe part gratitude…part guilt. That was a common feeling among many veterans I had talked with, and they often continued to feel its stab even decades later.

"Well," I said, trying to encourage him to continue, "I am glad you insisted on joining the Seabees. Otherwise, I might not be here." He looked over at me and almost smiled.

"You know," he said, "I got one deferment before I shipped out because I was workin' in the sawmill, and the country was in bad need of lumber for the war effort."

I waited, watching his face. His eyes seemed to be gazing back into the past…seeing faces and places that are now only ghosts.

"I was 19 when I volunteered in the summer for the Seabees," Dad said so suddenly it startled me. "I turned 20 in August, but they still hadn't called and so I thought I'd got lost in the shuffle. They finally called me in September and I went to Dallas to take an exam. I got Fireman First Class which was Third Class Petty Officer. They had done away with the Fireman rank…automation wiped that need out and they changed it to Machinist Mate 3rd class. So, I got to skip Seaman 1st, 2nd, and 3rd. That rating got me out of a lot of work detail like KP and other duties, but not at boot camp.

"There, it didn't matter what our rank was, we all had to serve. KP wasn't too bad; we entertained ourselves by making a game out of guessin' where the new guys were from. The boots were coming in so fast; we didn't have enough GI clothes, so they were still in their civilian clothes. We'd guess where they were from by the way they were dressed and how they looked, and we got pretty good at it.

"A lot of guys from the Southwest still had their cowboy boots on. One fella - I remember he was from Arkansas - he never put his regular boots on. He marched in his cowboy boots and everything."

"What did he look like?" I asked.

He thought a second. "He looked like somebody from Arkansas." We both got a chuckle out of that.

"We'd guess where they were from as they came through the chow line and then ask them where they were from when we served their food. We got most of them right. We never made any bets though because none of us had any money. Back then, we were all broke, we just did it for fun. Anyway…back to the story.

• •

"It was tough leaving home that first time. I'd never been away from home, except for huntin' and fishin'. It was sad, saying good-bye to the family, my mind was pretty well stirred up. There were about eight of us that met in Tyler where they put us on a bus and sent us to Troup to catch the train. It was about the first of October 1943 when me and the other men in that group caught the train out of Troup, Texas.

"We went to Camp Peary, Virginia by way of St. Louis, Cincinnati, and Richmond. I had never been out of East Texas until then. The trip took several days and along the route we picked up more men. We had no sleeper cars, so we had to try and sleep sittin' up in those crowded cars. When we arrived at the station in Richmond, we were met by Marines and a fleet of buses. When we got there, all we saw was Marines, even though we had volunteered for the Navy. The Marines were hollerin' at us to 'line up and get on the bus!'

"When we got to Camp Peary, the Marines told us there wasn't any place for us to sleep, so they marched us into a drill hall and told us we would sleep on the concrete floor. They had no clothes to issue, so we had to sleep in the clothes we had travelled in. We were all tired, dirty and hungry.

"Come chow time, they marched us to the chow hall, but by the time I got to where they were dishing out the food, I was sick from the odor. I got out quick and dumped my tray of food in the garbage. I bought some 5th Avenue candy bars at the ship's store and lived on candy bars for the next few days…until I could stomach the smell of the galley."

"Was the food really that bad?" I asked, incredulous that they were feeding young men preparing for war such awful stuff.

"It wasn't fit for pigs to eat." He shook his head. "Some of it didn't even look like any food I had ever seen...but, after a while you can get used to anything I suppose, and I started eatin' it just like everyone else."

"After two days had passed we were marched to a big parade ground with a speaker stand in the middle. There were about 1,200 of us standing at attention while they called roll and told us we would be issued our gear. We were still in our old clothes we had traveled in. The fellow calling out the names from the speaker stand was good. He didn't miss a single name. He pronounced even the Polish and Swedish names without hesitation and we had some lu-lus in our bunch.

"After roll call they sent us through a line and we took off our civilian clothes and wadded them up and they gave us a paper sack to put our clothes in. Then we were marched to the hair-cuttin' detail. It took about one minute for a man's hair to be sheared from his head. I had long thick black wavy hair and when the fellow that was cuttin' hair saw mine, he looked at me like ole Brer Fox looked at Brer Rabbit. He pressed those clippers to my head and said, 'If you want this pretty hair, hold out your hands!' He left me plumb bald."

Dad reached up and rubbed his hand through his still thick mane as though assuring himself it was still there. Even at 95 years old, he still has an impressive head of hair, although it has lost much of its youthful wave and is now as white as a snow-covered hill in January.

"From then on," he continued, "everyone with hair was our boss. We had been branded by our haircuts as low men on the totem pole.

"We made our way through the building and were issued our bedrolls, boots and gear. A lot of our clothes were leftovers from the 1930s CCC. They measured our feet to make sure we had the right fit. Men were coming in too fast to issue the proper dress. They gave us our shots while we were still naked. Then, after we had gotten all our gear, we were assigned to our barracks, which at Camp Peary were Quonset huts. There were about forty men to each hut.

"We were given a demonstration on how to roll our clothes and tie them. On our first inspection all our clothes had to be neatly rolled and laid out on the bunk. As the inspection officers came through, the head officer stopped in front of me and looked at my bedroll. I was six foot two inches and my bedroll was about five and a half feet long. He said, 'Get two or three guys to help you stretch that mattress – it's shorter than you are'. Right then, I knew I was in trouble if he was serious. You do not stretch a mattress! Hopefully, he was kidding, trying to lighten the mood and the stress of the moment." Dad chuckled.

"We couldn't go anywhere except the head, which was a building between the Quonset huts. Some of the guys would shoot dice in the head at night. I never felt led to gamble, always wanted to keep my money in my pocket." He patted his shirt pocket.

"We couldn't go to the chow hall by ourselves; we had to be marched there and back. They had a ship store on the way to the chow hall and we could dart in there and get a candy bar or something. That's how I got those 5th Avenue bars.

"Every mornin' we had to fall out for calisthenics while it was still dark. I thought I would be able to do push-ups, whatever they told us to do, without any problem. I was twenty years old and thought I was in top physical condition. To my astonishment, I could only do three push-ups. They had us exercise awhile, then go back and get ready for breakfast. After breakfast we'd march to the drill hall and then drill on the field. They were teaching us to follow cadence and stay in step.

"All of our trainers were Marines. Each one had a platoon to work with, and each one wanted his to be the best and they done a lot of hollerin' and carryin' on. What made it bad was when they would single out a person and call them out in front of everyone and chew them out because they couldn't stay in step. Getting about thirty guys of all lengths of legs and heights to step in time was a challenge, I'm sure. Our Marine instructor put us tall guys in the front and sloped size to the rear of the formation. Clifford Paxton and I were two of the tallest in that group and when we started out marching, we cut our eyes to the rear to see the short guys trottin' to keep up. The Marine instructor wasn't amused. He told us to cut our steps or he would put US in the rear! Clifford and I decided to slow down a bit."

He stopped talking at the memory of his best buddy and smiled, but his eyes betrayed his sadness. Clifford had passed away just a few months before, and I knew Dad missed him.

We sat in silence for several minutes watching the cows walking up the hill close to the house, following one of the many well-worn paths that wound through the pasture.

"Well," he continued, "we knew we had it rough at Camp Peary, but I didn't realize just how bad it was, compared to other boot camps, until I heard what Walter Winchell had to say about it. Winchell said, 'The Germans had their concentration camps and we had Camp Peary, Virginia!' There was a Congressional investigation of Camp Peary after I went through there.

"Of course, it wasn't all bad. At night we would fight the Civil War all over again." Dad grinned. "Every night the guys from the North would make jokes about the South and we would fire some back at them…the South always won because we outnumbered them about 2 to 1.

"Towards the end of November I was shipped out to Camp Endicott, Rhode Island. We were all glad to leave those freezing cold swamps of Virginia. When we got to Camp Endicott they had steam-heated barracks and we could walk freely from fence to fence without being hollered at. We could walk all over the camp when we weren't on duty. We felt like we had been let out of prison. I remember one hike we took, way out in the countryside. The leaves had mostly all fallen, but they were still colorful, a bright red and a golden yellow. They covered the road where we were hiking; it made for a pretty path.

"Anyway, Endicott is where I got put into the 31st battalion. The 31st had been on Bermuda doing some work for eleven months. They were nicknamed the Honeymoon Battalion. We got teased a little bit over that name."

He paused, looking thoughtful.

"You know, when we'd go take a shower, we would just pitch our wallets and pocket change on our bunks. I never remember anything being taken." He sighed. "Guess that wouldn't happen now…but we didn't think anything of it back then."

Dad looked over at me. "Did I ever tell you about joining the boxing team?"

"Yes," I said. "But you never went into much detail about it…except that you got your nose broken and got out after that."

"Well," he looked thoughtful for a minute and shifted in his chair. "It was after we got to Camp Endicott in Rhode Island from Camp Peary. Clifford and I found out that you could get out of all work detail if you joined the boxing team. So we thought it would be a great idea to be a boxer while we were at Endicott…bad decision!

"That boxing instructor worked us so hard I lost fifteen pounds in just a couple of weeks, and I was already boot-camp tough! He was from Syracuse, New York, a big fella and tough as a boot. He'd train us so hard our muscles would draw at night to where we couldn't hardly sleep. I remember I thought I was ready for anything…I could do thirty push-ups by then without any problem. Then our instructor hopped down on the floor, laid one arm across his back and started doing push-ups in rapid succession with ONE arm – Rocky style. It made me sick. I got as far away from him as I could to do my push-ups.

"They lined us up to catch the medicine ball after that. By the time I got my breath back it was time to catch it again. We caught it on our stomachs. He tried to kill us, at least that's what me and Clifford thought.

"Clifford and I decided we would have been better off if we had stayed with the work detail. It all worked out, because we both got hurt and off the team before the next competition, which was to be at Madison Square Garden in New York City, something to do with the Golden Gloves. I got my nose broken in practice and something happened to Clifford, but I can't remember what. We were both glad to be back on regular work detail after that, and gained back that weight we had lost pretty quick. The boxing team is where I got to know "Whitey" Knerr and George Young."

"When did you feel the most lonesome while you were at Endicott?" I asked.

"When it started getting close to Christmas," he said, and told a story about one of the guys in his outfit.

"My friend Martin DeVries told me they all went to see the movie *Holiday Inn* with Bing Crosby. It had just come out in the theatres. When they got back to the barracks, they were in their bunks trying to go to sleep, thinking about home, with the words to the song "White Christmas" still playing in their heads. That's when George Meeker, a boy from East Texas like me, started making the noise of a steam locomotive. It sounded so real. You could just imagine it going through the pine forests back home…some of the men were cryin' like babies. We all felt so homesick, being far away from home, especially as it got closer to Christmas."

"Did you ever get to go home for Christmas?" I asked, turning to look at his profile.

"Yes I did. My first leave from Camp Endicott was for Christmas. I went down to New York and caught the train there out of Grand Central, headed for St. Louis. Those old steam trains, they were all steam trains then, they could really eat up the ground. We asked the conductor, and he said, 'We're rocking between 95 and 100.' Lots of people don't know those steam engines could run that fast, but by the '40s they had gotten powerful. They were using everything that had wheels on it and could run on rails...shipping people and cargo. The whole nation was caught up in a whiz.

"Wasn't long, we were in Little Rock, Arkansas. After we got to Little Rock, we got on another train...and after being on that fast train, this one didn't only go slow, but it stopped at every little wide place in the road. It was crowded too, there were so many people on that train that there was no place to sit down. I had been gone a few months and was anxious to get home. That was the slowest train I've ever been on, we probably only made about thirty-five miles per hour between stops, but it was a good time to get to come home...at Christmas time. I got off the train in Jacksonville, but I don't remember how I got home. I know they were expectin' me, so I guess somebody was there to pick me up."

We both sat for a minute lost in thought. I finally glanced at my watch, then stood up and stretched, drawing the attention of a few cows that had worked their way up towards the barn just below the house.

"Let's take a break and get another cup of coffee," I suggested.

"Good idea," He agreed as he pulled himself out of his chair. "I don't want to get too stove up; I plan on going down to the home place later to check on the cows and maybe do a little fence cleanin'."

We refilled our cups and checked on Mama before we made it back out onto the patio. Dad settled into his chair and sipped on his coffee a moment.

"At Camp Endicott," he said, "we were close to the Air Base, and those fighter planes would come over our barracks and rattle the roof. Early in the mornin' they'd be takin' off and landing pretty steady.

"It was while I was at Endicott that I got an opportunity to ride in a B-25 or a Billy Mitchell, they called them. Anyway, they were a light two-motor bomber they used to fly patrol up and down the East Coast, and these two pilots asked me if I wanted to come along. Of course I was curious, so I took them up on the offer. It was a two hour patrol over a portion of the Eastern seaboard. I don't know how far we went, but it was a good ways. I was surprised to see all the little towns were lit up. None of them were blacked out...and we were in the middle of the war. The German submarines were sinkin' our merchant ships that were trying to get supplies to our troops in Europe.

"When their patrol time was up - I know they were just trying to scare me, but I didn't let on - they turned that plane and just fell out of the sky." Dad made a sideways diving motion with his hand. "I come plumb up off the seat, but I didn't make a sound. I didn't let them know that I was bothered by it…that was quite an experience." He smiled.

• •

"How long was it before you got to leave Camp Endicott after you got there from Peary?" I asked.

"After we got settled in at Endicott, we finally got a weekend pass to go into Providence. Joe Lowery and myself met these two attractive ladies and made a date with them. Me and my date went to the tap room, a sort of bar. She ordered us something to drink. Apparently it wasn't soda water because I got dizzy. I had never drank anything stronger than a Coke. I decided to find the bus station while I still could. I got on the bus back to the camp and the next thing I knew, the bus driver said, 'Hey, Mac! You have been riding this bus all night. This is my last trip, you better get off.'" Dad tried not to smile. "I never did know what happened to Joe.

"We had free transportation into town and back, which was nice, especially in the winter, but some of the guys would hitchhike. Our winter uniforms were all dark, so some of them started getting run over trying to get to town. They made a rule that we had to wear our white caps, even with our winter uniform, so we could be seen at night.

"At one point while we were in Rhode Island," he said, "we went out in the boondocks to work. It was the middle of winter by then and they marched us out from our steam-heated barracks right into the teeth of a blizzard. We marched five miles out to some unheated Quonset huts. They said it was 'advanced training'…I think they were just trying to see how much we could take; we like to have froze to death.

"When we got there, there wasn't a fence around the place, so we scattered like chickens. Our Skipper, or our Commander, got excited; he thought we were all runnin' off. One of the officers told him, 'they'll come back…they'll come back.' We did, but not before we explored the town close by…Westerly. It was about 40 miles or so from Endicott, right on the border with Connecticut. When we got back to the Quonset huts, we had to build fires with wood to heat them. They were nothing like our nice steam-heated barracks.

"Our building project was out there in a marsh, close to the waterfront. The ground was frozen and that's when I got acquainted with a jackhammer. We had to take those jackhammers and chisel out a hole to pour a foundation. We had to get through the frozen part and it was pretty deep there." Dad shook his head. "That was some hard work, but I was used to it so it didn't bother me the way it bothered some of them.

"One funny thing happened while we were working in that real cold weather. We all had to wear overshoes on top of our regular-issue brogan boots when we were outside to keep from getting frostbite. Those overshoes were pretty heavy, and when we pulled the overshoes off and our regular boots at the end of the day, our knees would fly up because our legs were used to picking up all that weight with every step." He smiled. "It was kind of comical-lookin' to see all of us high-steppin'."

He shifted in his chair again, drained the last of the coffee from his cup, and set it down beside his chair on the concrete patio. Then he placed his hands in his lap, right hand on top of the left, and stared off down in the pasture. He was squinting a little, as if trying to see something or someone in the shadows of the scattered trees.

"I remember one fella I met, named C.B. Choate," he said. "He had this lady friend who was covered with tattoos. That was a different thing back then. Not many women had tattoos and I'd never seen anything like that. I'd just come out of the cornfield. I wasn't used to things like that. Most of the fellas wouldn't have been caught out with her, but ole C.B., he was kinda dense anyway. That was some wild times."

Dad tried not to smile, but I could see the grin creeping up around his eyes and corners of his mouth. I couldn't help but laugh.

"One time," he said, "I went to a canteen that the town was running. This town was near a pilot training base where we were doing the construction job. There were a lot of girls there, but they were all chaperoned. It was a high class place where service men could dance with the girls and talk. I never was much of a dancer. The girls found out I was from Texas and they wanted to dance with me. I wouldn't dance, but when I sat down, a girl got on each knee. They wanted to teach me how to dance. I guess I should have been more sociable than I was. I always hated that I wasn't more sociable.

"Afterwards, we were walking some of the girls home and one of the girls said her sister - or it was a cousin, I can't remember - was a famous nightclub singer. She met us outside the canteen as we were leaving. There were about a half dozen of us guys walking those girls home and on the way, we passed over a bridge and one of the guys sat that girl who was a singer up on the railing of the bridge and she sang a song, just for us. She was the prettiest little thing. When we got back to camp, the guys told everybody about me having those girls sittin' on my knees, said I had more girls than I could handle. They got a laugh on me about that. Another time, I remember we went on liberty to a town called Artic."

Putting one of his hands on the arm of the folding chair, he wrapped his gnarled fingers over the end of the metal arm and began moving them slowly back and forth, a habit he has picked up since he got older. It's as if his hands are restless, unaccustomed to being still after a lifetime of constant work.

"Did you go into Providence much Dad?"

"We did. When we went into Providence on liberty, we'd sometimes spend the night, on the weekend, at a roomin' house. There were lots of them back then, not so many hotels like now. During the war a lot of people would rent out rooms to make a little extra money. Course, the houses were all shut up during the wintertime, so some of them didn't smell so good.

"I was short of money all the time because I sent most of mine back home to Mama. So I couldn't hang out in town very long. One old boy that lived in New York wanted me to go home with him for a weekend pass. He was an older fellow, about forty, and I think he had a daughter he wanted me to meet. Some of the Seabees were past forty years old. There were several guys in our outfit that were older. I was only twenty then... but I didn't go."

Dad had that faraway look in his eye - the look I've seen when he's recalling something he regrets. He sighed. "I guess I should have been more sociable," he said.

"Well, maybe I'm glad you didn't go to New York!" I exclaimed. He chuckled. I wondered if he ever thought about 'what if'? His life would have been completely different if he had married someone from another place, like New York. It was a strange thought to consider.

"It was a missed opportunity I guess. I could have seen more of the countryside. You could get on one of those fast trains and be down to New York City in nothing flat."

Not one to muse on regrets long, Dad sat up straighter in his chair, his face brightening.

"My favorite place to eat in Providence," he said, "was a place called Southern Fried Chicken. "You could get all you wanted to eat there for 65 cents and it was good. It was in downtown Providence. Most of the time we just rubber-necked around all over town, but I do remember walkin' to the state capitol. Mostly though, we just spent time lookin' the town over.

"Back then they had what they called 'tap rooms' underneath the hotels. A lot of the fellas…that's where they wanted to go, but even though I never developed a taste for the stuff, I went along with them. Never did see any fightin' or anything at those places. I guess the guys already had enough fightin' on their minds…in the Atlantic and Pacific.

"All of the young men had to wear their uniforms; it was a rule. If you saw a young man not in a uniform, somethin' was wrong with him. Those fellas were usually 4-Fs, unfit for service. There were a few that had deferments because of their jobs or family situations. I told you I had a deferment because I worked in a saw mill, but I didn't work long before I went into the service. Back then, if you saw a young man not in a uniform, that stood out…that was unusual because just about everyone was in the service.

"There were a few guys too that got into the service that probably never should have…like ole Steve. He was a guy that got thrown into our outfit from the brig while we were at Endicott. He was what we called a *cull* and was always in trouble. First time he went on liberty in Rhode Island, he came in drunk. Our barracks were long with a door at either end and a row of bunks on both sides with a clear space down the middle. Old Steve came up to one door and stood just inside and started yellin' and carryin' on and woke everybody up. Rogers had been named our barracks captain and he was supposed to keep order. Ole Rogers weighed about 220 pounds. He was from Fort Worth. His nickname was 'Buck,' Buck Rogers." Dad grinned. "Well, Ole Buck got up and went over to where Steve was at the door. Buck never said a word. He just hit him and knocked him plumb outside, turned around and went back to bed. That stopped that. Steve caused more trouble before the Seabees got their fill." He shook his head. "I just don't understand people like that."

Then he brushed a fly off his khakis and changed the subject.

"One of the fellas in our outfit was named Pester. I remember he was writin' a letter one day and we were called to chow hall. He left the letter where he'd been sittin' and I glanced at it as I left. I saw where he'd written, 'Mama…what are they doing to me!' I got a laugh out of that. I guess we all did kinda wonder what they were doin' to us - at least we wondered where we would end up. It's odd how some stuff will stick in your memory.

"While we were in Rhode Island at Endicott, we went several times to different parts of the state and even to surrounding states to do work projects. When we went to Connecticut on a work project, we would watch the pilots training at the nearby base. Some pilots would come in to land and hit the runway and bounce back up again. They had their trainers in there with them and they would tell them whether to shoot the gas to it and get up in the air again, or try to set it down. They were beginners. They were shoving them through that training because they were gettin' killed by the droves in them fighter planes. I never saw any of them crash while I was there. They might have, but I didn't see any then."

Dad turned to look at me. "Did you know they even trained pilots right here at home in Jacksonville?"

"What? I never heard about that."

He nodded. "Yep, they were going to Lon Morris College and they trained at the airport, which was where Nichols Industries is now...that was the old airport. When I was growin' up, we used to go to that airport and watch them jump out of airplanes with their parachutes - that was back in the late 1920s. They would put on air shows with all kinds of stunts. That was big entertainment back then.

"During the war years, Ole Herbert Coleman trained some of the pilots; he was an officer in the Navy. Frank Goodson, who ended up an ace pilot, decorated for bravery, went to Corpus Christi, Texas for his training and his boss was Herbert Coleman...both from little ole Jacksonville.

"Some of the fellas in our outfit liked to spend their time gamblin' and carousin'. I remember two old boys from Arkansas that liked to gamble. We had lights out at 9:00 every night at Camp Endicott, but every payday they would go out to the head where the lights would still be on and gamble. These boys talked long and slow and people thought they were ignorant, but they always walked away with all the money." He chuckled. "In fact, they made enough while we were at Endicott to buy them a new Plymouth. Their wives came up and they drove that new car back to Arkansas.

"When we first got out to California - while everything was still in a state of confusion when we first got there in the camp, they disappeared. That's how smart they were. They stayed in Arkansas until they got ready to come back, but when they got back they were in trouble. They prosecuted both of them. One of them was an officer, so he was just busted a rank, but the other guy was a seaman, so they locked him up in the brig.

"Later, when we shipped out to Hawaii, they were still serving time. The petty officer did something else - I don't know what - because he had to work off his sentence. When he finally got to Hawaii he had work detail on a truck. While we were hiking up and down those hills, he would pass by us on the truck, serving his time riding while we were hot-footin' it on those steep trails. His name was Hamner. I felt sorry for them for getting in trouble at first...but after seeing ole Hamner ride by in that truck on Hawaii, I felt a lot less charitable towards him."

Dad and I both laughed.

"My friend, Coon, from Louisiana, he could play the French harp. One time, when we were on liberty in Westerly, Coon started playin' his harp on the street corner. Soon, a crowd gathered to listen as he played and played. That was the only time I remember him playin' it in public like that. Mostly he just played in the barracks while we were restin' or fixin' to go to sleep. He could play real purdy.

"When we got a weekend pass, we could be gone from Saturday morning until Sunday afternoon. Lots of times though, we'd go Saturday morning and come back that evening, just so we didn't have to pay for a place to stay. Me and Coon spent a lot of Saturdays sight-seeing and just exploring the town or wherever we happened to be."

He paused as if he was going to say something else, but remained silent.

"Well," He leaned forward in his chair, pushed himself up with some effort and stood. "Guess I'd better get busy or I won't have time to work up an appetite before supper."

"Yep, you don't want to miss out on the fried chicken and gravy I'm planning on fixing," I said as we went back into the house. "Be careful working on that fence and don't forget to take you some water."

We placed our empty cups on the counter and I turned to leave.

"See you this evening, Dad."

"Ok and don't worry, I'll be fine," he said as he disappeared through the doorway to change into his work clothes.

Stepping off the curb, I hear a familiar voice. Across the way, in front of City Hall, stand Lawrence and Deliter Thorne. I recognize them right off. They look like they are about to get into their '38 Plymouth when I see Deliter stop and look my way. She puts her hand up to shade her eyes and says something to her husband who is fumbling with the keys.

Lawrence looks at her, then across the street at me. Crossing Commerce, they head straight towards me. I can hear Deliter's high-pitched voice rattling to Lawrence.

"I do believe that's Sis' younger brother, Stuart…looks like he just got off the 3:30 bus. Ask him if he needs a ride, Lawrence!"

The thin dark-haired man squints, tips his brown felt hat back as he approaches me and I hear him say, "I do believe you're right 'Liter. He's a might leaner than I remember, but there's no mistaking that head of hair."

Lawrence grins as he sticks out his hand. It feels good to have a real East Texas handshake.

"Welcome home, Son!" he says as he pumps my arm enthusiastically. I smile and then bow slightly and touch my forehead, nodding politely to Mrs. Thorne.

"You need a ride out to the home place?" Lawrence had read my mind.

"If it won't put you out too much…I can give you some gas money."

Lawrence waves me off as he heads back across the street, throwing the rear car door open for me, before climbing in behind the wheel. Adjusting my sea bag on my shoulder I hurry over to the car, hesitating at the open door.

"Are you sure it's no trouble?" I ask again.

"It's the least I can do," Lawrence assures me. "Besides, gas isn't as hard to come by now that the war is over."

Deliter starts talking as soon as I plant myself and my bag in the enormous backseat. She rattles on about who married who, where they live, and who's had babies, but I'm only half-listening. Cranking down the back window, I let the thick spring air rush over me. I lean into it, letting the warmth and familiar smells of the soil, fresh grass and honeysuckle fill my nostrils. I love those smells…they remind me of when I was a boy plowing behind old Kit and Dan. Deliter drones on and on. When Lawrence begins to slow for the turn onto the narrow red dirt lane to home, I ask him to let me out.

"I'll walk the rest of the way."

I meet Lawrence's eyes looking back at me in the rearview mirror. Deliter starts to insist that they take me the rest of the way, but Lawrence stops her.

"Deliter…he'll be fine."

Lawrence's eyes crinkle back at me in the mirror. I thank them and step out, pulling my bag with me. They wave and drive on, leaving me standing there, staring down the familiar country lane. I notice the small strands of bright green grass sprouting here and there between the ruts. I remember how I used to jump from clump to clump on the way to school, a contest to see how far I could go without stepping into the red mud. Funny the little things you remember sometimes.

Taking a look around before picking up my sea bag, I suck in one more lungful of the fresh air. When I start walking, I go purposefully slow. I want to savor every sight, every sound and every smell…I don't want to miss a thing. I can feel the anxiety slipping away with every red bird and crow call. I feel I am coming back…not just to home, but to myself. I will never be the same innocent country boy I once was, but I know I want to be a better man, one who makes a good life for himself and maybe someday, a family. I feel hope rising again and start to smile…yes indeed, I think to myself…there is life after war.

CHAPTER TWO

Today I helped Dad fix fence and check on the cows. He had his favorite coffee cup with him and was finishing his last cup of the morning as we climbed into Old Blue. All my growing up, whenever he finished with his coffee, if he was out in the pasture working, he would just turn his cup over and hang it on the nearest post or tree limb. When we were little, my brother and I would make a game of seeing who could find Dad's 'missing coffee cup'. Once, a cup was missing for almost six months before we found it. He had tucked it into the crotch of an old post oak about head high. If we hadn't been looking up to watch a red-headed woodpecker land, we would have never seen it.

When we reached the section of fence posts we needed to replace, Dad pulled Old Blue under a nearby sweetgum tree and we got out to survey the job.

Cup in hand, he squinted down the fence row and counted. "Looks like we'll need all we brought…maybe a few more."

I smiled when he drained the last of the coffee and turned the cup over onto a nearby cedar post. That cup was at least forty years old. When he accidentally broke the handle off of it as he put it into the kitchen sink one day, he couldn't bring himself to just throw it away, so he dug around in his shop and got some glass glue, the kind you use to re-glue a rearview mirror, and fixed that cup. That was over thirty years ago and the handle has stayed put. No doubt the ugliest cup in the cabinet, but he still uses it.

As we started to unload the tools and t-posts, I thought it was the perfect time to ask him more about his California days.

"To pass the time, why don't you tell me about what you did in California at Port Hueneme?" I asked as we started pulling steeples out of the ancient Western cedar posts – so time worn they were hardly bigger than broom handles. It's no wonder, since they have been here since the 1950s when Dad first built this stretch of fence.

He paused to scratch his forehead with a gloved hand.

"We left Providence and went to Grand Central Station in New York City to catch the troop train. In New York we had a little time before boarding the train, so we decided to do a little sight-seeing. It was cold that day, but we wanted to see the Empire State Building. We saw two little short fat policemen who had so many clothes on they looked plumb round. They stopped me and asked me where my pea coat (our heavy wool coat) was. I had left it at the train station with my other stuff. They started chastisin' me for not having my pea coat on. I told them 'Where I come from people dress like they want to, as long as it's decent.'

"Well, one grabbed one side of me and the other grabbed the other side and they started prophesying what they were gonna do to me…and I told them, 'There is a train leaving in a little bit and if I'm not on it, you guys will be in a heap of trouble.' And they turned me loose." Dad broke out in a big grin.

"From New York we went to Chicago, down to Amarillo, to Raton, New Mexico, across the Rockies and on to California. When we left Endicott headed for California, we got to see the whole length of America from coast to coast…the rolling plains, the farming country, cities, small towns, forests, deserts and mountains…we saw it all from that troop train.

"Over five days and nights, only stopping long enough to eat every once in a while, we got dirtier and dirtier. They were coal-burnin' trains, with no air-conditioning, so the windows were open.

"At our stops, the Red Cross handed out candy, cigarettes and other goodies. We left our dining car in Amarillo, so we didn't have to pull it over the mountains, I guess.

"At one of our stops in Raton, New Mexico, we got to go into this fancy hotel and eat at the restaurant there. Quick as the guys got in there some of them started ordering beers, but as soon as the officers got there they put a stop to that. They cut them off, told the hotel staff to not serve them anymore. Those that had them downed 'em pretty quick! I'm not a drinking man, but I always thought they should have let those fellas have a beer. After all we'd been through and were facing, it seemed like a small pleasure they could have allowed.

"That Raton hotel was a swanky place back then. When I went through there a few years ago it was all boarded up, but you could still peek through the glass doors and see that it was once an elegant hotel. It was kind of sad to see it just sittin' there…knowing it had seen so much history."

His voice trailed off and he seemed lost in thought for a minute; then the corners of his mouth turned up in a slight smile.

"I'd never seen the Rockies and at one point, we were winding our way through the mountains and I remember lookin' out the window and seeing our engine going the opposite direction. We were still going east and it was heading west! I also remember while we were going through the Rockies we went through a tunnel and the train filled with black smoke, soot and ashes. We were a dirty greasy bunch by the time we got to California! We didn't have a bath the entire trip.

"When we got to Port Hueneme, quick as we could, we got cleaned up and put some fresh uniforms on.

"From there, some of the guys went to San Clemente Island to do some work, and part of us went up to Ojai to build a golf course for the officers. I worked there running heavy equipment, moving dirt around, cuttin' trees, trimmin' limbs, stuff like that.

"The worst thing about my time in California was those ten-mile hikes with a full pack and carrying a 15 round semi-automatic carbine. The best thing about Port Hueneme was the liberties that we could take now and then. Of course, some of the guys figured out a way to "*extend*" their liberty privileges.

"One of our 1st Class Petty Officers had a wife that had come to see him before we shipped out. At the time, we were all under quarantine until we shipped out, and had our liberties all suspended. He came up with an ingenious plan to get off the base and see his family and helped a few of us enjoy a little more free time.

"The back gate to the base was nothing more than sand dunes and little else; the security wasn't extreme back then like it is now. The guys all put on dress clothes and then their work clothes over them. The officer would march them to the back gate and holler, 'Work detail!' and march them out the gate and down the beach. When they were out of sight of the gate, they pulled off their work clothes and buried them in the sand. Off they went on liberty! They would gather back at a set time, put their work clothes back on and the officer marched them back in!

"They couldn't get away with somethin' like that now, with all the cameras and surveillance, but back then was a simpler, less complicated time."

Dad stopped to lean against a pull post for a minute and catch his breath. He chuckled.

"It was at Port Hueneme that one of the funniest things happened to one of our guys. A Yankee named Curly Welch had the biggest blonde head of curls you've ever seen and he took quite a bit of razzin' over it. Some of the guys decided it was time for Curly to get a trim, so they grabbed him and held him down and whacked off all those bushy curls.

"Just as they finished trimming Curly, Lieutenant Ermilio, a stocky Italian, was spied walking towards them and they scattered like squirrels. Poor Welch was still on the ground, feeling of his shorn head when, without thinking, he called out to the Lieutenant, 'Meatballs! They cut off me courls!!' No one had ever dared to call him Meatballs to his face, but I guess Curly was so upset he didn't even think about it! We all had a good laugh about that…except maybe Welch." We both laughed.

"Lots of the guys seemed to have nick-names," I said as I reached for a box of metal ties we used to secure the barbed wire to the t-posts. "Did you have a nick-name too?"

He pulled off his left glove and shook it, releasing a little shower of Bahia grass seeds. "Yeah, we gave most of the fellas nicknames or shortened their names. I don't know why. We just did. They called me Tex…not too original, but easy to remember," he drawled as he pulled his glove back on.

"Well," I said, "it suits you." That made me think about how far away he had been and how that nickname must have reminded him of home.

"How long were you in California?" I pulled open a new box of post ties. "Did you ever get to come home again?"

Dad stretched his back, then picked up the bucket we were using to hold the metal ties and passed it to me so I could refill it.

"I believe we were at Port Hueneme from sometime in April till the 2nd of October 1944. Around the first of June the whole outfit got a leave to go home. It was about ten days, but a lot of that time was spent traveling. It took me a good two or three days to get home by train. I rode the train all the way from Oxnard, California to Jacksonville. Back then trains would stop in all the little towns, not just the bigger cities.

"When I got home, it was tomato picking and shipping season. The packing sheds were runnin'. It was a nice time to be home. In a way though, it was strange. All of the guys I knew were gone in the service, so I didn't do much, just visited family and looked the town over. That June leave was the last time I saw Grandma Josephine Young, Mama's mother.

Someone had contacted Mama and told her she was real sick, so Mama got someone to go and pick her up and bring her to our house. Grandma lived down between Moss Hill and Batson in a little house out in the country. I can still see her lyin' there in that northwest bedroom of the old house. She died while I was on Iwo Jima…so did President Roosevelt."

He paused for a minute as if searching his memory.

"I did spend some time visiting your Mama while I was home on leave." He smiled. "I remember the night before I left to go back to California; we went to the picture show. I walked her home. We didn't see each other for almost two years after that night."

He paused again, pliers poised in mid-air, then shook his head.

"Before I knew it, it was time to go back to California. I'd only been home about a week. When I boarded the train in Jacksonville to head out west, I recognized an old boy on the train that day named Lightsey. He was in my outfit. He was older than me and married to a girl named Christine; I went to school with her. Lightsey sat down by me and was reading a letter. He started cryin' and got to where he couldn't read it no more and handed me the letter and said, 'Read this…it's the damndest thing I ever seen.'

"His wife was tellin' him she was leaving him. A Dear John letter and here he was going overseas, not knowing if he was coming back. It was sad: I felt so sorry for him. It upset him so that by the time we got to El Paso, he was sick. I got off the train and found a shore patrol and told them they had to do something with Lightsey. He didn't have a dime, so he borrowed five dollars from me. It was all the money I had, but I didn't need it since we were going back to the base. I didn't want him to be dead broke. Later on, he showed up at the base, but I don't remember him shipping out with us. That was a sad episode."

"Poor fella," I had to agree. "Sometimes it makes you wonder why people do the things they do."

Dad shook his head. "I can still see him sittin' there, tears runnin' down his face…his whole world crumbling with the stroke of a pen."

We worked in silence for a while before he continued.

"Speaking of pens," he said. "Did you know my Mama, your grandma, taught herself to read and write so she could write letters to her sons in the service?"

I knew that Ma (what all of us grandkids called her) was an amazing woman, but had no idea she had accomplished such a feat all on her own.

"Wow! She was determined to keep in touch, wasn't she?" I handed him the pliers.

"Yep. There were three of us boys in the service during the mid-forties, Floyd, myself and Monroe. Floyd was with MacArthur when he returned to the Philippines. Monroe went in 1946 and was sent to Germany and then later to Korea. I know she wrote to all of us. I got letters from Mama pretty regular, but I didn't keep any of them. Now, I wish I had. But then, it was hard going…I just didn't think about it till later. The only letter I still have is one that Grandma sent me."

Dad shook his head. I didn't say anything, but I knew those letters would have been such a treasure and a comfort to him, especially now that Ma is gone. I wondered again…what makes us do the things we do?

Neither one of us said anything for a long while. We just worked, lost in our own thoughts. Close by, in a scope of woods across the fence, I could hear a woodpecker. It sounded like a miniature jackhammer echoing through the creek bottom and beyond. The air was still cool even though it was after midday, and I could smell the faint perfume of a late blooming honeysuckle somewhere in the tangle of briars nearby.

Suddenly, he cleared his throat.

"While I was in California, I was assigned to duty at our base jail…on suicide watch. If guys got into bad trouble, sometimes they were watched to make sure nothing happened to them. That was one of my jobs while I was there. In Rhode Island I served on the Shore Patrol for a while. There were nine of us in that unit. Your Mama found a picture of us in one of her photograph boxes the other day. Remind me and I'll show it to you when we get back to the house."

I nodded. I love seeing those old photos of Mom and Dad and their lives before I came on the scene. It reminded me that they were people too, with their own hopes and dreams - not just Mom and Dad.

I could tell he was getting tired, so I suggested that after we finished this section we call it quits. He let out a big 'whew' and nodded.

We worked for a while enjoying the woodpecker's determined rat-tat-tat in the woods nearby.

"There must be some mighty tasty bugs in that old pine for him to carry on so." Dad said with a grin.

I agreed. "He must be one hungry bird."

Making our way down the fence row, we replaced a whole section of those original cedar posts. As I handed him the tools and he twisted the wire ties on and tightened them down, we developed a rhythm that made the work go swiftly. Soon, we had fixed that part of the fence almost as good as new. After we had surveyed our morning's work and picked up the bits of leftover barbed wire and began loading up the tools, he started talking again.

"While we were out there in California, we were more or less markin' time until we shipped out. They did different stuff to keep us busy. One thing I did, they assigned me to this *old* painter…he was probably thirty years old." Dad smirked.

At ninety-five, that thirty-year-old must seem mighty young now, I thought to myself as we finished loading and climbed back into Old Blue to head home. We slammed the cranky truck doors so hard that we startled a pair of doves that rose and whirred off towards the woods.

"We worked paintin' the window screens - the wooden frames had to be painted back then. We painted all the window screens on the base, there must have been several hundred. It kept us busy for a while. That painter was in our outfit. There were a lot of experienced painters, carpenters, and other tradesman in the Seabees. That's how I got my third-class petty officer's rating, from my work with the sawmill and runnin' equipment.

"Another job I had was helping this officer write all the numbers of the carbines. They assigned me to help him read those numbers. Some of them, we had a hard time seeing, they were pretty small. He would say, 'Tell me the number, lad! Tell me the number!'" He laughed. "There were about eleven hundred of them to go through and I don't know why we had to write all those down, but I guess we had to turn the carbines in when we went on leave. Then, when we got back, we had to reassign them.

"Another time, me and another fellow were assigned to this older officer who was experimenting on packaging for loading supplies aboard ship while at Ojai. We were using steel bands to band the supplies together so they could be moved by forklift and cranes to load onto the ships. I worked for a while with him.

"Well…'nough about that!" Dad said as we drove through the pasture up the hill to the house - his way of saying he was through talking about the subject for the day. I just smiled to myself and enjoyed the bumpy ride home.

We unloaded the tools and put them in the small shed near the house. I could tell he was ready for a shower and to prop up his feet for a while. "I'm going to check on Mom." I told him as he started back towards his room. "I'll see you tomorrow."

"Ok, you have a good evening."

I went into the Big Room to see if Mama was awake. She was in her chair working on her Wheel of Fortune puzzle book when I came in.

"Hey Mom!" I said as I plopped into the rocker opposite her recliner. "What are you working on?"

She peered over her glasses at me. "Do you want to try and guess this one?"

Mama likes to challenge us to see if we can guess the answer to a puzzle early, because she usually can. She was always good at Wheel of Fortune puzzles on the television. She hasn't watched the show for years, but when she did, she would always guess the answer before anyone else. Now she just works the puzzles in her Wheel of Fortune word search books, and is just as good at those. When she showed me the puzzle with only three letters filled in, I guessed it right away, much to her surprise.

"San Francisco Bay!" I announced with a grin.

"Boy," Mom said, "you're pretty good."

"Well, I learned from the best."

Smiling, she put the book on her lap. "How's the fence building going?"

"We're all through for the day, Dad needed to get cleaned up and rest awhile."

She nodded. "He's probably worn out. His foot has been swelling some too." Mama looked concerned.

"He will be good as new tomorrow," I said. "At least, that's what he always says."

Mama laughed. "Yeah, that's what he says, but he's not able to bounce back like he used to. Still…he does pretty good for ninety-something."

"Mama?" I hesitated for a second and then plunged ahead. "Do you remember going with Dad to the movies the night before he left for California?"

"What brought that up?"

"Oh…Dad was just telling me about when he came home on leave in June of '44 and he mentioned he saw you then."

She nodded. "Yes, we went to the picture show and he walked me home that night." She smiled. "I'll never forget it. We first got acquainted when we were living on old Doc Travis' place, just across the valley from the McAnally farm. We had to walk across their farm and through their yard to get to school at Enterprise. When I first really noticed Stuart, he was older and had already quit school. He was working at a saw mill in a scope of woods not far from the school. Sometimes when I'd pass by, we'd wave at each other. We'd known each other for several years before he left for the war. I was only sixteen the night he took me to the show." Mom looked wistful as she recounted the story, momentarily lost in the memory. I cleared my throat, bringing her attention back to the present. She looked like she was considering her words before she continued.

"Like I was saying, the night we went to the show, he walked me home. It was a beautiful night…the moon was shining…we were standing there talking and all at once, he just put his arms around me and kissed me – it was my first kiss." Mom smiled. "I sure hated to see him go.

"I wrote to him for months after he left and he wrote back…then one day," Mom's voice cracked. "I got a Dear John letter from him. I was so hurt. I didn't write to him anymore after that."

I was surprised. Dad had never mentioned writing a Dear John letter to her…even when he told me about Lightsey and his letter.

"Well," I exclaimed, "I'll have to ask him about that!"

She shook her head. "I was young and didn't understand why he would do such a thing, but now, I realize he was just trying to protect me. He didn't know if he would make it back. He was heading into combat… he just didn't want me to get my hopes up and then something happen."

"That's Dad." I nodded. "Always looking out for others."

Just then the phone rang. It was my sister Melanie checking on Mom and Dad and asking about their day, so I waved good-bye and let myself out the back door.

All the way home I thought about Mom and Dad and being young during war time. It must have been hard, but when they looked back, they usually talked about the good things. Life is like that, I thought, no matter what time you're born into…good things and bad things, it's all part of it. You don't get to choose what your times are as Gandalf says in *The Lord of the Rings*; you only get to decide what you do with them.

Shivering under the stack of heavy quilts, I can see my breath in the little bit of light coming from the woodstove. It must be one of the coldest winters ever. I heard Dad talking to some of the old-timers when we went to town today to get some tobacco for his pipe. They were saying we might even get some snow.

I smile and squirm further down into the quilt pile. Snow for Christmas... wouldn't that be something.

Dad clears his throat real quiet-like and I turn my head towards the sound. I can just see his profile in the almost dark room. I hear him tapping his pipe against his hand and then see him reach inside his jacket pocket for his bundle of Twist. Dad always has two things with him, his pipe and that Twist. He sometimes chews and dips but, in the winter time, his favorite is his pipe.

He let me hold it once. It's "a thing of beauty" he calls it. The bowl is a dark reddish wood, carved like a fierce-looking bull's head complete with horns. It is perfectly shaped, a miniature wonder.

I watch as Dad pulls a piece off the Twist and puts the rest back in his pocket. I know his routine by memory and even in the near dark room, I can see him poke the tobacco plug into the bull's head and reach for a match. When he lights the match, it throws the whole room into shadows and movement.

In a few seconds, I can see the glow from his pipe and catch a whiff of the tobacco smoke as it drifts towards the cracks in the opposite wall. The north wind is 'getting up' as Dad would say and starting to whistle through the cracks beside my bed.

In the quiet, watching Dad, listening to the soft sounds of my sleeping brothers and sisters, I feel happy. Not because I expect anything for Christmas, we don't get toys... maybe an orange or an apple if Mama can manage. No, I am happy because I feel loved and safe. I reckon that is the best gift of all.

Suddenly, I feel something cold on my cheek. I turn my head towards the wall and see some tiny white flecks shimmer in the faint light. They keep coming, faster and faster. Finally, I realize it is snowing! I want to jump up and shout, but I don't want to wake the others, especially Mama, since she's been cooking all day for Christmas dinner tomorrow.

So I smile and close my eyes, keeping the secret to myself till morning. I can't wait for Christmas…

Chapter Three

I had a quiet day planned for a change and hoped Dad did too because I wanted to hear more about his adventures in California.

Dad's memory has always amazed me. He can remember just about every job he has ever done as a painter and he can recall names and faces of people everyone else has long forgotten. He is a real stickler for the truth and always prefaces a story by giving credit, if it is second-hand. When he talks about his time in the service, he always tells us who said what or told him about something he didn't see for himself. Sometimes, he can't remember a name or a place, but that isn't very often. When he tells a story, almost everything he relates can be corroborated by others and by the old worn CB book he keeps in a wooden box in his closet.

I asked him once about that box, not really expecting a great story, but I got one.

"Who made that old box you keep all your papers in from WW II?" I asked one foggy spring morning.

He was at the kitchen table, working on a piece of raisin pie my sister Sherry had made the day before. Raisin pie is his favorite...and mine. I thought everyone knew about raisin pie, but when I'd mention my favorite pie, people would look like I'd told them I like dirt pie. They had apparently never experienced Mama's delectable recipe. It is a combination of buttery, flaky crust, plump tender raisins in a lemony filling that would rival any chef's efforts. Dad and I adore raisin pie! I watched and waited as he scraped the last bite of pie off his saucer, took a sip of coffee, and leaned back in his chair.

"It was a fella from Fruitland, Maryland - Larry A. Brown. He was in my Company...Company C." Dad grinned. "Brown was a nice-lookin' man; he delivered milk before going into the service. You know, like they used to, delivered milk from door to door. Well, old Brown entertained us with stories about women coming to the door in their nightgowns to get their milk." He gave me a sideways glance. "Don't know if they were all true," he added, "but they were entertaining.

"Brown made that box on Iwo. I asked him where he got the wood, hinges and stuff to make it. He said he 'procured it'." Dad smiled. "Brown had more points than me, so he got to go home from Iwo. He didn't want to take the box, so he gave it to me. I toted that box to Japan and all the way back home. It was heavy too." He laughed. "Of course, I was a lot younger then, so I could heft quite a bit."

I nodded and laughed. "That *would* make it easier. What are your plans for the day?" I asked as I watched him finish his cup of coffee.

"I think I'll just be lazy today and enjoy the cooler weather."

"Good!" I got a cup from the cabinet and poured myself some coffee. "I want you to tell me more about California and Port Hueneme."

"Well," Dad said, "I guess talking won't be too strenuous. What do you want to know?"

"First, I want you to tell me about the Dear John letter you wrote Mama." I gave him a stern look. Leaning forward, I said, "You didn't tell me about that little piece of information."

Dad looked kind of sheepish and cut his eyes around at me. "What in the world brought that up?"

"You mentioned about taking Mama to the picture show yesterday, but you didn't elaborate, so I asked her about it."

"What did she tell you?" he asked.

I just laughed. "No way," I said. "You tell me your version and then I'll tell you what she said." He tried not to smile, but I could tell he was getting a kick out of it.

"Not much to tell," he said. "We went to the picture show and I walked her home."

"That's it?" I looked at him and cocked an eyebrow.

His eyes crinkled. "That's all I'm tellin'."

I laughed. "Mom was more forth-coming with her version, but I can tell you remember what happened, so we'll just skip to the Dear John letter part."

The smile left his eyes and he looked almost hurt. "She wrote the sweetest letters and I wrote back…for a while, but I could tell she was hoping I'd come back and at that point none of us knew if we would come home. I hated to do it, but I was fixin' to be facing combat and I didn't want her to be waiting on me, expecting me, and me not come home." Dad shook his head. "She never wrote back anymore after that. I felt bad, but I wanted her to be free to date other fellas and not feel tied down."

He stared out the picture window by the table.

"I never had feelings for any other girl except your Mama – not like I did for her. I went with other girls, but never felt that way about any of them."

I smiled at my Dad, his face weathered with time, but his eyes as bright and shining as the twenty-year-old that fell in love with a skinny sixteen-year-old, dark-haired girl under a Texas summer moon. Every child should be lucky enough to have the security such an enduring love offers; I am one of the lucky ones.

I swallowed a lump that had suddenly formed in my throat and got up and went to the stove to pour me some more coffee. Neither one of us said anything for a few minutes. I finally sat back down and waited for him to break the silence.

He suggested we go out to the patio and sit, so we moved out there and got comfortable. Only then did he start talking again.

"While out in California, we got to go to a baseball game, see Hollywood and Los Angeles, Santa Monica and the fancy Hollywood USO where movie stars would take turns serving and talking to the servicemen.

"But it wasn't all fun and games though; we had training to do and some of it was pretty tough. One of the things I remember was an LCI advanced training we were doing just off the coastline…we were practicing exiting the LCI via the ramp at the end. Our Lieutenant Ermilio was mad about having to exit into the water. He was short and when he went into the water a wave swept him under. He was fit to be tied!

"The worst part was, we all had to march back 13 miles to camp, wet with salt water. We were all raw and blistered from our packs and where our clothes rubbed us as we marched. It was one of the most miserable hikes we took during training."

"My word!" I said. "That must have been awful."

Dad shook his head. "I thought it was pretty dumb at the time, but we didn't have any say so…you just did what you were told. Of course, some of them, like ole Steve, I told you about earlier…they just did what they wanted.

"I remember one time he took a boat without permission, and when he was on liberty, if he saw a car with a key in it, he'd get in and just drive off and drive it until it ran out of gas. It's hard to imagine now, but people left their keys in the car, would even leave them runnin' to dash into a store. Way on up into the 1960s and 70s, you could still do that in Jacksonville. Mostly nobody bothered them, but ole Steve couldn't resist it, I guess. They finally got tired of foolin' with him and gave him a dishonorable discharge and sent him home."

I shook my head and wondered silently what made some people so prone to mischief. I turned to face him. "Could you tell me more about going out on the town while you were in California? That sounded like fun."

He glanced over at me and took another sip of coffee.

"Well…every other week or so, we'd get liberty for the weekend. We'd leave on Friday afternoon and we didn't have to be back until roll call on Monday mornin'. We usually went to Los Angeles, Santa Monica, or Hollywood. One time, a bunch of us went down on busses to see a football game; the University of Southern California was playing, I think it was Washington State, in the Los Angeles Coliseum. That was the first football game I had ever seen.

"Another time we went to see Roy Rogers' Rodeo at the Coliseum. I remember that Dale Evans was about 19 years old; that was before her and Roy married. I remember her singing. It was kind of a dude rodeo, nothing like our rodeos in Texas." He smiled.

"They named the celebrities that were watchin' the rodeo the day we were there. I remember one was Hoot Gibson. Bill Elliott, a cowboy that was in the movies a lot, was Roy Rogers' ringmaster in the show. I met Bill one time back in the thirties when he came to Jacksonville. One of his movies was showin' at the theater and back then movie stars would travel around to the theaters to advertise their movies."

"I never knew you met him!" I said. "What else did you do?"

"One time the three of us, Cecil "Coon" Watkins, myself and Clarence Peterson went down to see *Ken Murray's Blackouts of '44 play*. We got there and they said there were no seats, but they told us if we waited around till after the show started, there might be some 'no shows' and we could get some seats. Turns out, we got seats on the third row from the stage!

"Marie Wilson was his leading lady for the show. They had an intermission and there were three WAVs, Navy girls, sittin' to our left.

When they got up durin' the break, ole Ken Murray told us, 'You sailors keep your seats!'" Dad grinned.

"We could also go to the Hollywood Canteen and get free coffee and donuts and stuff like that. It was nice. Different movie stars would be in charge of the canteen, takin' turns donatin' their time. I remember one time I was there, Bette Davis was in charge."

My eyes got wide. " That is so cool!"

He chuckled. "Of course, to a young buck like me, she was an older lady, so I wasn't impressed. I saw her, but I didn't talk to her or anything... it wasn't a big deal to us guys. We appreciated what they were doin'; of course, they had the money to make it a real nice place to hang out.

"They also had a place where you could spend the night, a kind of dormitory. They took real good care of the servicemen. We went there several times while we were stationed in California because it was handy."

"Things were a lot different back then," I said. "I can't imagine the movie stars of today doing anything like that now."

Dad shook his head. "No...I guess it wouldn't be the same; times have changed."

He grew silent, so I used the pause to go in and check on Mom.

"I want to hear more when I get back," I called over my shoulder as I stepped inside the house.

When I peeked into the Big Room, Mom was sitting at her little card table setting out a new puzzle to work on. She turned and smiled.

"Well, hello there. I was wondering if you were going to come in here and say anything."

Mom sometimes feels left out when Dad and I get on one of our story-telling/sharing binges.

"Oh Mama," I said, "you know I always come in here to see you."

I walked over to the table and helped her turn over the puzzle pieces.

We visited awhile about family news and she filled me in on the latest letter from her niece, Wanda. She writes the best letters, several pages long and always full of funny happenings and the antics of her numerous cats.

After about thirty minutes, we had most of the border of the puzzle together, so I went back out to see if Dad was still on the patio. Sometimes he gets tired of waiting on me and leaves for the barn or goes down to the home place to check on the cows. He doesn't sit still too long, so I was pleasantly surprised when I found him still out on the patio.

"You ready to tell me some more of your adventures?" I handed him a glass of water as I sat down beside him. "Passed by the kitchen and thought you might like some water."

He took the glass, thanked me and took a long draw. "I guess I needed that." He set the glass down and folded his hands in his lap, one on top of the other before he started his story again.

"There was a pleasure pier, they called it, at Santa Monica. They had a roller coaster out over the water. I remember me and Coon were together and we were with these two ladies. The one I was with was a…um…full-bodied girl." Dad cut his eyes around at me and smirked.

"We rode the roller coaster and when we went over the top of that coaster and it started down, it looked like we were headed into a hole that was gonna to be too small for us to get through, so we all ducked down. When we folded up to tuck our heads, the girl I was with bursted her bra. It was funny!" We both laughed. "We had lots of fun while we were there at Port Hueneme.

"While I was in Ojai workin' on that golf course and helpin' with those shipping containers, I met a girl. We went out a few times, and then I was sent back to Oxnard, Port Hueneme." Dad shifted in his seat and looked over at me. "I've never told anyone this story before."

"Really?" I raised my eyebrows.

"Really," He said. "Guess I was a little embarrassed.

"Anyway, I went back up to Ojai to see this girl on a Saturday. The last bus to the base ran about 10:00 or 10:30 that night and I missed it. There was no place to stay in Ojai - it was just a small village back then - so I took off walkin'. It was 13 miles back to the base.

"When the sun goes down out there it starts gettin' cool, especially in the higher elevations and Ojai was kind of up in the mountains. Well, the further I went, the colder I got. I saw this hayfield out by the side of the road and the hay wasn't baled…it was just piled in stacks."

He motioned with his hands as he described the scene. "I went out into the field and dug into a pile of hay and finally got warm enough to go to sleep.

"When I woke up the next day, the sun was out and it was warmin' right up, but I had hay all over that wool uniform. I had to pick all that hay off before I got out in the road…I didn't want to scare anybody." His mouth twitched, like he wanted to smile.

"When I got back to the road, I discovered there was a house across the road from the hayfield. I didn't even see it the night before. By the time

I had walked a little ways, these people in a car stopped and picked me up and gave me a ride back to the base. I sure was glad to get back…that was an experience I didn't relish at all. I never did go back up to Ojai to see that girl."

"Did you ever meet any other girls you liked while you were there?" I asked.

He thought for a moment.

"There was this one fella in our outfit - his parents must have been well off because he had a pretty fancy car. He was going out with these sisters and he couldn't make up his mind which one he wanted to date, so he talked me into going on a date with the other sister. I said I would, but I wasn't too happy. The girl wasn't very nice-lookin' and I was pretty picky." Dad grinned kind of sheepishly.

"Anyway, their daddy ran a filling station and they had to paste all of the gas ration stamps into the books before we could go. We went for a ride up towards Santa Barbara and around. One of them invited her girlfriend to go with us and that kind of messed the conversation up. There we were in the backseat with her friend. It wasn't much fun. I never did go with him again anywhere, so I don't know how it worked out.

"Me and Clarence Peterson went to a USO dance one time in Santa Monica. We met some girls we both liked while we were there. Peterson met a gal from Ardmore, Oklahoma, her name was Mary Garvey. And I met an Italian beauty named Lou Renzo. Both were living in L.A. at the time. I walked Lou home from the dance that night. We stopped along the way and ate at a little café with her sister and her boyfriend. I remember because he got up and left me with the ticket. I didn't appreciate that. I was pretty short on money. I was upset with him…he was a con artist and he knew exactly what he was doin'.

"Anyway, I paid the ticket and walked Lou home. She lived in a two-story house with her mother and her sister. I don't know what had happened to her father. I remember we ate watermelon when we got to her house and we talked for a while. It was already pretty late, so she invited me to spend the night on the couch downstairs, but I had already made arrangements to stay at a YMCA not far from there, so I told her good night.

"Peterson and I were talking about the girls later and decided we both liked them and that it wouldn't be fair to get involved right then, since we were due to be shipped out in just a few weeks. We made a pact to not see them anymore. I kept my end of the bargain, but I found out later that Peterson didn't."

Dad didn't elaborate at the time, so I filed that question away for later. It was getting on towards noon, so I told him I'd have to go home and get some things done. I waved to him as he walked out to his shop to "piddle around," as he calls it, then went inside to tell Mom bye.

• •

While I washed the dishes that evening after supper, my mind wandered back to Dad's story about Lou Renzo, the girl he met at the USO dance.

What if he had married her, I wondered? She was a dark-haired beauty like Mama. Dad seemed to favor dark hair…maybe because they seem more sincere and down to earth…or maybe, I thought, he just likes dark hair.

As I dried the last of the dishes and began putting them away, I wondered what might have been, how his life might have been different. What if he had kept seeing Lou? As I put the last of the silverware in the drawer, I decided we were both probably glad things had turned out the way they did.

The view from these plow handles isn't the best in the world. Old Kit's and Dan's backsides are about all I can see looking straight ahead, so I sometimes let my mind...and my eyes wander to the creek and the cool shade of the sprawling red oak tree I pass on my round to the west side of the field.

I try to keep the rows straight. I can hear Dad's voice reminding me to keep my mind on my business, so I take a deep breath and focus on the ridge the plow is making beneath my bare feet.

The late spring weather has been cool, and with my old straw hat keeping the sun off my head, I have to say I'm enjoying this year's plowing. Dad said we'd plant the usual – cotton, corn, a few tomatoes, some beans and peas, maybe some watermelons for the 4th of July.

My mouth waters thinking about those juicy red melons. They are a rare treat and when we make a crop, us boys make ourselves nearly sick eating as much of them as we can hold. Mama puts up all the vegetables and fruit she can for the winter, and the rest we give to others who are not as well off as we are.

We are poor...I know from the way the other folks dress when I get to go to town with Dad to sell the cotton. Our clothes are clean, but patched and sometimes too small...or too big.

I don't feel poor, especially when Dad pays me for picking cotton. When he goes to the gin he always lets us kids ride in the wagon on top of the cotton sacks. That is one of the best days of the year. Dad gives us a nickel for helping with the picking and the loading. A nickel is a lot of money, so while I'm working, I usually spend a considerable amount of time thinking about how to use it.

Last year, instead of giving me the usual nickel, Dad gave me a whole quarter, an unexpected treat. Before we left for the gin, I overheard Mama say she was out of sugar and didn't have the money to buy any. Staring at that big quarter in my hand, I made up my mind to buy Mama the sugar she needed instead of the candy and marbles I was going to buy. The look on her face when I gave her that sack of sugar was worth the sacrifice. I felt ten feet tall!

A diving blue jay scolds me out of my thoughts and I realize too late that I've made a bobble off my straight and narrow furrow. I vow to spend the rest of the afternoon with my mind on my job.

Come evening, I see Dad walking out to inspect my progress. I wait for him to say something about the crooked row, but he just stands there surveying the field, nods to me and says as he turns towards the house, "Supper's on." Relieved, I cluck to Old Kit and Dan to head home for the day and promise myself that tomorrow, I will do better.

Chapter Four

"Knock! Knock!" I called out as I opened the back door. Seeing Dad at the stove through the half glass door, I knew it must be coffee time.

"Hey," he called back when he saw me. "What are you up to?"

"Well, I came over to have coffee, if that's what you're doing."

"Yeah…I could use a little company." He set the stainless pot off the fire. "Your mama's been feeling poorly, so we didn't get our porch time this mornin' like we usually do. She's in her room, might ought'a check on her."

I nodded, then walked back to her room and gently knocked on the door.

"Mama? You all right?"

"What?" Came the weak reply, then a shuffling sound and finally, the door opened.

Mama looked smaller than usual - bent over with her long silver braid slightly askew. Her face looked forlorn and tired.

"Wish you felt better, Mama," I said. "Would you like some coffee? Dad's making his afternoon pot."

She made a face. "No, I don't think so; my stomach doesn't feel too good."

I could see over her shoulder and couldn't help but notice the pile of chocolate candy wrappers and the open cheese cracker box by her easy chair. No, I thought; I don't believe I'd feel too good either, if I had eaten all that stuff. But I didn't say that.

"Hope you get to feeling better," I said.

"Thank you, hon. Always remember I love you."

Mama has taken to saying that every time one of us kids leaves. I suppose it's her way to insure a final pleasant memory, in case of her sudden departure from this present world…or ours. Mama never did like to leave anything to chance. She feels like it is tempting fate to not cover all your bases. Maybe she's right, I thought.

"Love you too, Mama," I said.

As I turned to head back to the kitchen, I heard her door shut quietly and the TV come on. I knew she would probably be asleep shortly. Mama has always been a night owl, staying up half the night. She naps a lot during the day, but if you catch her, she'll deny it every time.

She doesn't always eat on the same schedule as Dad either; oftentimes they will eat hours apart and alone. Since it doesn't seem to bother either one of them, we have adjusted to this new schedule. Of course at family gatherings and on holidays, we all eat together like we used to when we were growing up.

Dad has never let much ruffle him, even when Mom gives him a long list of honey-dos. The only thing that ever really bothers him these days is not being able to do what he used to do.

Just the other day he was telling me how *little* he could accomplish.

"I ran the chainsaw for only four hours before I gave out and had to rest," he said. "Then I loaded the wood for the fireplace and hayed the cows, but I had to lay in the floor awhile when I got home…to recover."

I just looked at him and laughed. "Dad! You **are** going on ninety-six years old! And you're complaining because you can't do what you did at 90!? Most people are moldering in the grave by the time they reach your age."

His mouth twitched a little before breaking into a grin.

"I guess I do pretty good…for an old man," he said.

Growing old has been at times difficult for Dad, but really, he didn't start to slow down until he passed ninety. When you've always been strong enough to do whatever your mind determined to do, then the day comes when you tackle something and you struggle to do it…that has to be hard.

I walked into the kitchen and saw that he had set our two cups on the worn Formica counter. He fixed his coffee as usual with a little sugar and cream, then put the cream back in the icebox, as he and Mom still call the refrigerator.

"Want to sit outside?" I asked.

"Sure, it's a fine evening'. Not too chilly yet, but we got another norther due tomorrow," he said as he picked up his cup and headed out the back door.

"Well, that's alright with me," I declared as I followed him. "I like cooler weather, and we'll get some good fireplace time in."

The folding aluminum chairs still sat where we'd left them the last time we were outside, so we just brushed a few gold and brown leaves off the seats, settled into them, and sipped our coffee.

While we looked out over the hills and ponds of the pasture that stretched out behind the house, Dad let out a small sigh. "Never gets old… does it?"

"No sir," I replied. "I've been looking at it all my life and I expect the rest of my life…and I never get tired of it."

We sat and enjoyed the quiet, watching the cows and calves as they huddled around the hay rings and munched away, hearing the occasional crow or hawk call as they echoed across the monochromatic landscape of late fall.

This has become our routine - afternoon coffee on the patio, weather permitting, by the fireplace in winter, or the sunroom when it's blistering hot or raining. When my sisters visit, we all enjoy an early morning cup of coffee together before we begin our day, except Melanie… she's a tea person. After a struggle, we have come to grips with the fact that you can enjoy tea about as much as coffee! When he can, my brother stops by and joins us. These are happy days. We know they are precious and few, so we treasure them all the more.

"Dad?" I reluctantly broke the silence. "Would you finish telling me about your time in California? I want to hear about your trip to Hawaii too, and what it was like to leave the states."

He got that familiar look on his face, like he was turning back the years in his mind, searching for the right information and spot to begin his story. His eyebrows scrunched together, then he wiped the back of his hand over his mouth and cleared his throat. I polished off the last of my coffee, turned my chair slightly to see him better, and waited.

"Well…let's see. I told you about all the places we got to go and some of the things we did already. I guess I'm down to where it was time for us to leave the states," he said more to himself, I thought, than to me.

"When it got close to when we were going to ship out, they quarantined us. It cramped the style of a lot of those boys because their wives had traveled out there to be with them. Bud McEwen's wife was one of those who had come out to California. She was a classy-lookin' lady. She worked in the office there at the base. She found out where we were going to ship out to and swore us to secrecy.

"We were headed to Hawaii. She found out through the information coming across her desk. She could have gotten in a lot of trouble, but we didn't say a word. All of the wives and families that came out there had to make plans to either stay or make their way back home. It was a hard time for a lot of folks.

"One of my buddies, Jerry Seiberlich, a Russian from Oakland - we called him Si - wanted to go back one last time to visit his family, so we cooked up a plan between us and several other guys to answer for him at roll call so he could make the trip home. I remember he said his Dad owned a shoe store. We took a chance getting caught, but we got by with it. He got to spend a nice long week with his folks. Afterwards, we met in Santa Barbara at a restaurant and Seiberlich bought us all dinner as a thank you. He really appreciated that. We were all a little anxious about shipping out, so if we could help each other out like that, we did. Seiberlich and I stayed buddies all the way through our time in Japan, after the war.

"On October 2, 1944, we boarded ship to leave the states…it was a sea of sea bags and white caps as we winded our way up those gangplanks. It was a peculiar feeling knowing you were leaving your homeland and not knowing if you would return. We were heading into the unknown. You realize it may be a one-way trip…and for a lot of the guys on that ship, it was.

"Martin told me about one of our guys, Meeker, who had gone AWOL the week before we shipped out. He had wanted to see his new bride home to East Texas before we left. Martin said when we were loading up, he heard someone hollerin' 'Wait for me!' It was Meeker, running to catch up to him in line. He didn't want to be left behind. He wanted to be with his battalion wherever they went."

Dad got that far off look again as he paused.

"Meeker never saw the States again…he was killed on Iwo Jima. He came home to his new bride in a box.

"He wasn't the only one…Clifford was standing next to Whitey Knerr as we left port. Lookin' over the rail of the ship as the land got smaller and smaller, Whitey said, 'I sure hate to leave the States…I don't think I'll come back.' Whitey's words were prophetic, he was killed on Iwo.

"When we started boarding the ship to head to Hawaii, I remember reaching the top of the stairs leading up to the ship, and stopping and turning around for one last look at the mainland…and home. I remember thinking it might be the last time I see it. It was a sad feeling. I'll never forget it. It was an eerie feeling when you're heading out towards that wide open ocean. None of us knew if we would come back, and so many didn't."

He shook his head. I wanted to cry. The picture of my young Dad in his sailor uniform taking that final look at his homeland made me ache for all the young men who never came home. I'd seen photographs of the troop ships leaving the states in Dad's CB book and often wondered how many of those young men in the photographs never saw their homeland again. It was a sobering thought.

I watched Dad's face for a minute to try and discern his mood, but like other times when he speaks of the war, it was hard to read. I've learned at those times…it's best to be patient. So, I waited. I didn't have to wait long.

"The ship that took us to Hawaii was a converted Merchant Marine ship called the Aqua Prince. Boy, was that misleading! It was a real tub. It was fitted with air-conditioning, but some guys would take their knives and cut holes in the air duct so the air would come out on them, it really shorted everybody down the line. We nearly smothered down there.

"There were several thousand men aboard, and it was crowded above and below deck. A lot of the guys played cards and dice to pass the time, but I was as sick as a horse with sea-sickness the whole way over…a lot of the guys were, even though the ocean was as smooth as glass for our trip to Hawaii, nothing more than a gentle roll.

"One night, they put me on guard duty. I didn't know what I was guarding way out there in the middle of the ocean, but there I was, so sick I could hardly stand up. Lookin' back, I think that part of that was just plain old worry and nerves, because later, on the way back home, the ship I was on went through a typhoon and did everything but break in two, and nobody got sick.

"I also remember they played the World Series over the loudspeakers on the way to Hawaii, and sometimes we could see whales jumpin' out of the water near the ship.

"It was six long days before we reached Hilo and they started dividing us up into our Companies and our assignments. Some of the guys were assigned to load and unload the ships for the invasion. At that time, we didn't know but what they might be sending us to an invasion of Japan. We had no idea till much later where we would be going - it was top secret, but that didn't keep us from speculatin'.

"When we got to Hawaii, they gave all of us a swimming test. In the Navy, they are gonna make sure you can swim. I don't know if they do that in the Army or not, but if you were in the Navy, you had to pass the swimming test. They put a buoy way off out there where the water was deep. If you could swim out there and back, well…you could swim. I thought, if I fail this, I'll get to do some more swimming in that nice warm water, so…I just couldn't make it back." Dad cut his eyes around at me. "But I was sharp enough to get to where I could touch bottom because it gradually got deeper. They culled me and I had to stay down there for a week, swimming in that nice warm water.

He smiled. "It was good for me; I got to where I could relax in the water and turn over on my back and float. I soon joined the others and found out what everyone was doing.

"It was still pretty warm there, even though it was October. When we got to Camp Tarawa, they issued us another blanket. I thought, well, we spent a winter in Rhode Island and here they are issuing us another blanket? But before mornin', I knew why. In the daytime it was warm, but you'd freeze at night. It was pretty high elevation for an island, about 2,600 feet. It was mountainous, rough terrain.

"Some of our guys were assigned to build a camp for what they thought were some WACs, but it turned out to be war dogs. I think those fellas were mighty disappointed." He chuckled.

"We were sent to Camp Tarawa to train with live ammunition with the 5th Marine Division. They loaded us onto flat cars on a little narrow gauge train. They had a pole run along the sides of the cars to hold onto. We rode it up the side of the island to the end of the track. When we got to the end of the track, we were picked up by Marines in trucks and hauled to Camp Tarawa. Along the way we got a good look at the island. The sugar cane looked like it was growin' wild, but they had these chutes or flumes, up and down the mountainsides to float the sugar cane down when they harvested it. They were built way up high in places and closer to the ground in others. They had to give it so much fall so the water could float the cane down to the sugar mill near the beach. Those flumes looked like elevated miniature railroad tracks from a distance.

"It was rough mountainous terrain," he said motioning up and down with his hand, "lots of canyons and steep mountainsides. At one point, our engine came unhooked. It was already headed out over a tall bridge and while they were hookin' it back up, I got off the flat car and cut me a stalk of sugar cane. It was that old hard Japanese cane, we called it. It wasn't very good to chew, not like the sugar cane back home. When we got to the end of the railway, they put the engine on a turn table and it was balanced so that two men could turn it and they could hook it back up. That was the end of the line and it's where we got off the train and onto the trucks.

"Our camp was on this big ranch that was on the island. At that time, it was the largest ranch in the world, next to the King Ranch. Where we were, there was nothing but volcanic ash; nothing would grow on it but cactus. We used those big cactus for target practice. There were a few eucalyptus trees, but not much else. Where the land stuck out of that volcanic ash, the Bermuda grass grew almost knee deep. The cowboys that worked on the ranch then were New Zealanders.

"We had some pretty tough hikes while we were training. That was where I spied ole Hamner, that boy from Arkansas riding by in a truck, while I was hoofin' it with a full pack and my part of the 60mm mortar. I wanted to yank him out of that truck." Dad made a face. "Didn't look like much of a punishment to me." he said and started to smile.

"Every day, there was a mist of rain that came over the island, but by 10:00 a.m. the dust would be blowin'. We would nearly freeze at night. I realized then why they issued us those extra blankets. I couldn't imagine we'd need 'em on a tropical island, but we sure did.

"One time, on a hike, we saw a cloud move over on top of Mauna Loa or Mauna Kea; I'm not sure which one. That cloud sat there for a while and when it moved, the top of the peak was covered with snow. I didn't know it could snow in Hawaii. I watched, and every day the same thing happened. Gradually, every time a cloud moved on, the snow line got lower and lower down the mountainside. Our water supply came straight off the mountain, cold as ice.

"Our showers were unheated, so no one lingered. There were no lines at the showers. You knew when someone was in there too…you could hear them squall out like a panther when that ice water would hit them!" He laughed.

"For those of us from the South and Midwest, Hawaii was like being on a different planet. It was the first encounter for many of us with palm and coconut trees and tropical weather. While we knew that things were gonna get pretty serious very soon, we tried to keep our spirits up and not worry…at least for now.

"It was while I was in Hawaii, at Camp Tarawa, that I found out Ole Pete had been lyin' to me. Pete, myself, a fella named Armstrong, and a big guy from Arkansas - I remember he had a scar on his stomach, I guess from an operation - we were all in the same tent. A funny thing about Armstrong was, one day I asked him to *fetch* somethin' and he told me there was no such word as 'fetch.' I told him, 'There most certainly is!' but I don't think he believed me." Dad smiled.

"Anyway, we all shared a tent there at Tarawa. Our tents were big canvas ones, about fourteen by fourteen feet, up on wooden platforms and there were five men to a tent. Two on each side, end to end, and one at the back, I can't remember who that fifth fella was in our tent.

"Anyway, one day a guy came into our tent lookin' for Pete. I told him he had the right tent, but Pete wasn't there, but he'd be back shortly. That fella was Mary Garvey's brother, and he told me that Mary and Pete were engaged and planned to get married when he got home! I couldn't believe Pete had lied to me. Then, I thought about those last few weeks we were in California. Pete didn't go on liberty with me after that USO dance where we met Mary and Lou. He told me that he was goin' down to San Diego to see his brother…when in fact; he was still seeing Mary, even after we made a pact to not see them anymore.

"When Pete got back, I jumped him out about it. I told him, 'I've considered you my buddy all these years and you've been lyin' to me. I don't want to speak to you anymore.' And I didn't for a long while. If I wanted to tell Pete something, I'd tell one of the other fellas to tell him and Pete would do the same. That went on and on till one day we were on our way to chow hall and Pete turned to me and said, 'Tex! Let's knock this stuff off…you're running me crazy!' Well…I finally forgave him and we went on being friends - good thing too, because we ended up in a foxhole together on Iwo Jima." Dad laughed. "Old Pete didn't lie to me any more…not that I know of anyway."

"I'll bet he thought twice about it, for sure." I smiled. "Did you and Clifford Paxton spend any time together on Hawaii?" I asked.

"We did." Dad nodded. "Clifford and I spent our off time explorin' the island, at least part of it.

"Clifford told me about him and a few other guys when they were down at the beach wonderin' how they could get some of those coconuts down. Those coconut palms were at least 30 feet tall or more.

"Clifford said they tried to throw rocks at them to knock the coconuts off and they managed to get one or two down; then they spied a little boy, maybe 10 or 12 years old, watchin' them. They asked him how to get those coconuts down. He said, 'Climb the tree.' The guys all laughed at him and asked him if he could get them a couple of coconuts. He said, 'Yes, for 10 cents each.' Clifford told him okay and up that tree he went!

"Clifford said he threw two down and came scootin' down that tree like a monkey. They got such a kick out of watchin' him shimmy up that tree that they gave him 10 cents each time and sent him back for more. They soon ran out of dimes, or Clifford said they would have sent him up again.

"Wonder if they were worth the trouble...and all those dimes?" I joked.

"You bet. I ate some coconuts while I was there. The milk and meat were delicious...but it was a battle to get the husks off! You know...Clifford sent some of those coconuts home to his mother. Last time we talked, he told me he still had them. Funny the things you hang onto."

I laughed. "Wonder if they are still any good?"

Dad shrugged his shoulders. "I doubt it, but those husks were awfully thick.

"Another thing we enjoyed about being on the island was gettin' fresh-baked coconut pies from the islanders for ten cents each." He made a circle with his fingers and thumbs. "They were about six inches across and worth every penny. One day I ate so many of those pies I got a stomachache. I didn't want a coconut pie for a while after that." He chuckled.

"The training on the island though was serious. I was a gunner on a 60mm mortar. We would practice by hittin' the cactus that dotted the landscape. When we marched, we had to take the gun apart and divide it up among three men. We carried it along with our regular gear. There were a lot of days we had to crawl under live ammo and throw hand grenades."

A dark shadow seemed to pass over Dad's face. He paused.

"What is it?" I asked.

He reached up to run his hand over his face as if to wipe something off. "Not every man was ready for that part. We had some guys get scared and jumped up into the live fire...I don't think their families ever knew what happened." He sighed. "Maybe that was for the best.

"I'll never forget one of the guys, a Marine, froze during hand grenade practice. We had a wall of sandbags about two feet high for protection during practice. We were supposed to follow orders. The grenades had a safety pin in them and you caught ahold of the lever that went down the sides of our hand grenades. The Japanese hand grenades were different - they had a detonator in the nose and if that nose didn't hit, they didn't explode. Anyway, you pulled the pin on ours and had your hand around the lever and when you threw it, the lever flew off and when it landed, it exploded.

"We were behind the sandbags and the officer gave us the order to 'pull the pins!' and then gave us the order to throw. Of course, we knew to duck down behind the sandbags, and then the officer would holler for us to 'stay put.' He was watching us, and one guy, a Marine, pulled the pin and froze. The officer had to run over and pry that live grenade from his hand and throw it.

"If a person was killed during training, their folks never did know how they were killed, unless they had a buddy that told them. They were just told that they were killed in the line of duty. One of our demolition experts had a shell go off when he was takin' it apart – it messed him up, but it didn't kill him. One guy we nick-named Jersey Jerk shot a hole in his tent. We still don't know how he managed that.

"We all knew that things were building up to something big; we just didn't know what. I overheard one of the supply officers who had been on the *USS Missoula* say that before the ship left the islands they had loaded cargo on the weather deck…cans of napalm, flame-thrower fuel, and boxes of body bags." Dad stared off into the distance. "I tried not to think too much about that. The officer said he didn't know where the *Missoula* was headed, but it must be *hot*.

"When it got close to Christmas, we found out we would ship out soon and the officers threw a battalion-wide barbeque on the beach. We had a good time that evening, but I'm sure in the back of everyone's minds was this was our last chance to relax and have fun.

"Some of the fellas decided to start throwin' one another out in the water. They would gang up on somebody, haul them out a ways and toss them into the waves. Everyone had a laugh until they tried to pick on Big Moe. Mohare was from a little town just southwest of Jacksonville in East Texas. His 6 foot 6 inch frame and heavy-build made him a giant among the 31st men.

"The first three guys that tried to pick him up and haul him out got tossed like sticks into the water. More and more of them joined the effort, but ole Big Moe tossed every one of them out into the water! They never did get him down! We had a good laugh over that. I remember me and Clifford sittin' on the beach watchin' the fun and laughin'. It would be a long time before we would laugh again."

Dad looked sad and happy at the same time. It was a look he reserved for good memories that made him sad. As I watched his face, it occurred to me that as I've gotten older, I can better understand those types of expressions. Because I too have memories that are happy ones, but they make me long for a lost time…or for a person who's no longer here. I don't know if young people can understand it…maybe it comes with years. His voice broke through my reverie.

"We got one more leave before we headed to Saipan," Dad continued. "We went to Honolulu, docked and had one day's liberty. We saw some of the results of the Japanese attack on Pearl Harbor; you could still see the Arizona sunk out in the harbor… although a lot of debris had been cleaned up by then. I remember takin' a look around Honolulu. There were thousands of sailors on the main drag of town; it was a sea of white uniforms.

"Of course, it looked different then; none of the big hotels and things was there. I went into a hardware store and bought a whet rock to sharpen my GI knife. I still have it, but somebody broke it. I can still use it though."

"I remember you using that rock ever since I was a kid," I said, "but I didn't know where you got it."

How like Dad, I thought, to buy something practical when others were probably spending their money on one last drink or on a girl. I could imagine him hefting the whet rock in his hand, mulling over his purchase, frugal in his thinking as always.

"In fact," he said, as he rose with surprising swiftness from his chair, "I think I'll go get it and sharpen my pocket knife."

He headed down to his shop, but I decided to stay put and enjoy the view. The pasture this time of year always reminds me of a brown carpet, streaked with gold from the lengthening rays of the evening sun. A lone cricket chirped occasionally from the woodpile by the shed, as if to remind me that winter would soon be here.

Pretty soon, Dad reappeared, and as he settled into the chair beside me, I leaned over to take a look at the rock. It was indeed battered on one corner, but still useful. It's hard to believe he's kept up with it all through the war and still has it after seventy-plus years.

"You were saying, about being in Honolulu," I prompted.

He fished his Old Timer knife out of his front-right pants pocket, opened it and started sweeping the blade across the whet rock.

"At the upper end of town, where I was," he said, "you could see the whole main drag of Honolulu. I was watchin' these guys taking pictures from that vantage point, lookin' across that sea of white caps. I watched them for a while and noticed they took this guy's picture, then neglected to get his address so they could send the picture home to his family. I thought, what's going on here?

"I went down to the police station and told them about it. They said, 'We're watching them.' They were spies. At least it didn't surprise the police. I'm glad they were watchin' them.

"There were two things that tipped me off. First, they were young guys, military age and second, they failed to get that sailor's address. That was a dumb mistake. They should have sent some older guys to do that job; they wouldn't have stood out so much. They probably didn't even have any film in that camera. They were definitely German spies.

"I didn't know until much later that not everyone left when we did. Several of our 31st bunch, including me, left Hawaii, headed to Saipan, around the first of January in a convoy of eight LSTs. The rest came over on a troop ship much later. For us, it was slow going. The LSTs could only do about seven or eight knots an hour, top speed. The troop ships could make about 17 or 18 knots an hour, so that tells you how slow we were. It was a miserable ride.

"We still didn't know our final destination when we left Hawaii. We wouldn't learn that until after we reached Saipan. On those LSTs, it was hot and muggy with nowhere to go. There wasn't enough room or air to sleep below, so we spent most of our time topside."

"What did the LSTs look like?" I asked, trying to picture them in my mind.

"Well, the LSTs didn't have a name like the ships. They just had a number and they were about as big as a football field. The doors opened in the front, like the Higgins boats, so they could hit the beach and unload the heavy equipment, tanks and such. The bunk areas were on either side of the LSTs, and the deck was for the heavy equipment we were carrying. The chow hall was at the stern, the aft, or the back of the LST. We were carrying half-tracks, ammunition and such for whatever invasion we were headed for… we still didn't know.

"We only had two sub-chasers for protection. We were slow, so we were more or less sittin' ducks. If an enemy torpedo had hit us with all that ammo on board, we'd still be going up!" Dad quipped. "But somehow, we never seemed to attract any attention. It was quiet; we didn't hear a thing, except those old LST pontoons rockin' against the sides…day and night. We did get a couple of scares onboard though. Those big horns would go off 'Whoop! Whoop!' and boy, we would scramble around. We never knew if they were just drills or the real thing.

"We got along pretty good for being in such close quarters. There were about thirty or forty of us plus the crew. Of course, the crew thought they should always be first in the chow line since we were just along for the ride, but that didn't go over too good with us when they started cuttin' in line.

"In the military, it's a no-no to cut into a formed line, unless you were on duty; then you could go ahead of the rest. There got to be a hassle in the chow hall over fellas cuttin' in line, so they appointed me as the kind of MP of the chow line. It was my job to keep the line straight and under control. Some of the little fellas would bow up at me, but I'd just say, 'To the end of the line or over the side!'" He smirked. "Back then I was over 200 pounds…I didn't have any trouble out of anyone.

"Our chief cook was an older fella, and he would let me come back to the galley and he'd make toast for me. That toast sure was good."

Dad has always loved toast; now I knew why. He has it with breakfast, sometimes as snack or with his coffee in the afternoon.

"We didn't have toast growin' up," he said as he continued making sweeping motions with the knife on the whet rock. "We ate biscuits for the most part, or cornbread, so I enjoyed havin' that toast…it was a real treat.

"On our way to Saipan, we had to stop off at the Marshall Islands to repair one of the LST's engines. It was an uninhabited island just above sea level with beautiful white sand and palm trees. That island looked like what I had thought Hawaii would look like, a country boy's version of a real tropical island. It made for a nice break from the monotony of the LSTs.

"I got permission to go ashore and took advantage of the crystal clear water surroundin' the island. When I put my head under the water, I wasn't sure about opening my eyes, because at home, all you could see was reddish murky haze, but when I opened my eyes underwater there, I could see real good. The salt water stung a little at first, but it didn't take me long to get used to it. I had quite a time swimmin' around in that clear warm water. I could see colorful fish thirty to forty feet down. I really enjoyed that."

"I'll bet you did," I said. "Quite a change from Mud Creek."

He nodded.

"When did you finally make it to Saipan?"

"We arrived in Saipan about the 13th of February. That's where we changed over from the LSTs to the troop ship and joined the rest of the 31st and the 5th Marine Division.

"Boarding the ships in Saipan was a dangerous thing. We got off the LST we had traveled on across the Pacific, and onto the Higgins boats to get to the troop ship. The seas were choppy and we had to go up on this wide net of ropes from the Higgins boats to get onto the deck of the ship. Several men were hurt trying to climb the nets…some of them were crushed against the ship by the Higgins boats as they tried to get on the ropes.

"Our gourd-headed officer had forgotten to get all our paperwork in order, so we couldn't move away from LST until it got straightened out. We were jammed in that Higgins boat like sardines in a can and could hardly move. Unfortunately, I was at the rear, right where the LST's diesel exhausts were bellowin' fumes. I got gassed and was so sick I didn't know if I could make it up those ropes when we finally got to the side of the troop ship.

"The guys on the ship wanted to drag me up in a net, but I was determined to board under my own power. When I finally got to the top, I was so sick I just fell over the railin' and onto the deck. I couldn't even stand up for a while. Not a good beginning for the last leg of my trip."

I shook my head. "I can't believe that officer didn't know better. That could have affected you for life."

Dad cut his eyes around at me and made a crazy face. "How do you know it didn't?"

I laughed so hard I got the hiccups.

Floyd, Monroe and myself like to go down to the creek and hunt for frogs or fish whenever we can, so when Dad tells us we are finished with chores this morning, we light out for the back of the place and our favorite spot on the creek. I am the oldest and fastest, so I beat them both to the big white oak and am already skittering down the banks towards the cool water when they catch up with me.

It's hot and muggy, a typical East Texas summer day, so the cool creek water feels good on my legs and callused feet. I watch as Floyd and Monroe jump into the water nearby splashing and hollering for all they are worth. We all laugh and jump around, enjoying our freedom from supervision and from work.

Being the oldest, I always feel responsible for the younger kids, even at play, even more so for Floyd. He has what Mama says is a 'tangled tongue' and that makes him a target for some of the boys that we go to school with at Enterprise.

I'm glad we only go to school after the crops are laid by in the fall. Being shut up inside is torture for me and my brothers. It's especially hard on Floyd because the other boys poke fun at him because of the way he talks.

I've had plenty of scraps on his behalf. When I come home from school after one of those episodes, Mama usually scolds me for getting in trouble and for fighting, and then hugs me real tight. I know it's her way of saying 'thank you' for protecting Floyd. We all look out for him in our own way.

Today, Floyd and Monroe want to catch some frogs, so I watch as they go along the highest part of the creek bank and look into the holes just above the water line, trying to see if they can spy some big fat bullfrogs. I am content to just wade here in the cool muddy water and watch.

Everything is going fine until Floyd decides to put his hand into one of those holes…I guess he saw something and thought it was a frog, but it ain't…it's a big ole water moccasin.

It all happens so fast…Floyd hollers. I catch a glimpse of the snake's huge head as he pulls his hand out of the hole. I know I have to move fast, so I grab Floyd's other hand, pulling him along and holler at Monroe to come on and we all light out for home as fast as we can move.

I keep looking down at Floyd's hand as we run for home; I see it start to swell, then his arm, too.

When we get about halfway home, Floyd slows down and says, "I can't run anymore…I feel sick…"

He stumbles. I pick him up, even though he's almost as big as me and I start to run. My heart is pounding against my chest like a hammer and my ears are ringing.

"Please don't die…please don't die…" I keep saying over and over in my mind.

I see Mama out in the yard as we top the rise above the house. Monroe passes me.

"I'll tell Mama what's happened," he says, running towards home.

When I get there with Floyd, his arm is already twice its normal size. I look at Mama, scared to death of what might happen to Floyd. Just then, Uncle Joe pulls into the driveway and Mama waves at him to come and help.

Uncle Joe and Mama get Floyd into the car and take off for town and the doctor. Monroe and I are left standing in the dirt driveway watching them disappear in a cloud of red dust and I'm praying my brother won't die.

CHAPTER FIVE

"I hear your son," I said as Dad and I sat out on the back patio late one afternoon. It was one of those unexpected, but always welcome autumn days in East Texas where the temperature gets up to almost seventy degrees, making it seem like spring - for a few hours.

He tilted his head and turned his ear towards the road.

I listened as the unmistakable sound of the paint truck my Dad had driven for years shifted to climb the steep hill to the driveway, slowed down, and turned in. By then I could tell Dad had heard it, because his face lit up.

"Sounds like my boy!" he said.

We waited while Shaun pulled to a stop and got out. I stood up to greet him with a hug as he walked over to join us. He hugged Dad and sat down in the chair opposite him.

"How did your day go?" Dad asked.

"Aww…pretty good, I guess," Shaun said. "I'm almost through with that big job out at the lake. I've got to stain the deck and do some touch-ups."

He filled Dad in on all he was doing. Dad always loves to hear about Shaun's work and the people he's seen in town. Since Shaun took over the painting business that Dad had built, he has kept Dad's stellar reputation intact, drawing rave reviews from his customers. He still does work for some of the same people Dad painted for when he was still working. They've known our family for years and often tell me how happy they are that Shaun is painting like his father. It is obvious that Dad enjoys staying in touch with what's going on. A lot of the people he knew while he was a painting contractor have 'gone on to their reward,' as he is fond of saying, but he still likes to talk about the painting business with his boy.

I kept quiet and watched the two of them. The years seemed to melt away as Dad listened and advised my brother. He is so proud of his only son; well, I thought…the only living one.

As I looked at Shaun's chiseled profile, his high cheekbones, aquiline nose and strong jaw, I marveled at the resemblance to Dad at that age. The older Shaun gets, the more he favors him. I couldn't help wondering what my brother Josh would have looked like now…if he had lived to see his fifties. I wondered if he would look more like Dad today, too.

The sun had worked its way past the roof and was peering around the west end of the house while they visited. I looked over at Brother and saw that a ray of sunlight rested on his hair, highlighting the silver glittering in his still mostly-brown flattop. I smiled. Not everyone can pull off a hairstyle that has been mostly abandoned for over a half century, but Brother can and does. He was blessed with Dad's thick full head of hair and has the exact same hairline. Now, at almost fifty, Brother's trademark haircut will soon be the same color as Dad's - those shiny sparkles are just the beginning.

Shaun got up, stretched and told us his plans for tomorrow. He patted Dad's shoulder as he walked behind his chair and gave me a hug good-bye before he headed for home.

"Have a good evening," Dad called, as Shaun got in his truck.

"Sure glad he's got plenty of work," he said as he struggled to rise from his chair. "This time of year can be slow."

I gathered our cups before heading inside. The sun had now disappeared behind the row of pines at the edge of the pasture fence, and it had gotten noticeably cooler. Dad shivered as he walked into the shadow of the house.

"It's coolin' down," he said. "Time to build a fire."

As we walked into the kitchen, I asked him if he had time to tell me about when he left Saipan."

He stopped and leaned on the counter for a second.

"Having a dizzy spell, Pop?" I asked.

"Oh, I'll be alright; it comes and goes."

"If you don't feel like talking, it can wait."

"No, no. I'll be okay. I don't have anything else to do this evening… unless your mama takes a notion to play Mexican train."

I laughed. "I wouldn't want to stand in the way of her game."

"Well," he said, "we can sit in front of the fire and talk until she decides, I guess."

"Okay. I'm going to get me a glass of water and I'll meet you in the Big Room. You want a glass too?"

"Sure, I probably need another glass before bed."

After I poured our glasses of water, I head through the dining room and down the three steps into the Big Room.

After a couple of minutes, Dad appeared in the doorway. He had changed into the house shoes with the fleece lining that my sister Sherry got him for Christmas last year.

Mom was in her room adjacent to the Big Room. She had dubbed it with that name since Dad built it onto the house in 1984. If anyone called it the living room, Mom always corrected them. Judging from the voices coming from her bedroom, she was watching *Pride and Prejudice* for the umpteenth time. She's tried to get Dad to watch it with her - six hours' worth – but so far, he's managed to avoid it.

Dad added a couple of logs to the fire, and got a good blaze going before taking his seat in the recliner. He admired the flames for a moment, and then began his story where he had left off earlier.

"After we all got on the troop ships in Saipan, the Marines boarded the LSTs that we had rode over from Hawaii. They were preparing for a landing. We didn't know for certain where it would be, but we had a pretty good idea.

"Some of our guys, the ones who traveled across the Pacific on the troop ships from Hawaii, found out where we were going on the trip over, but the ones of us on the LSTs…we didn't find out until after we were underway from Saipan that we were headed for Iwo Jima. We left Saipan and headed north…Iwo was about six hundred miles away. It would be a pretty quick trip.

"Tokyo Rose told us on the way over that we were headed for Iwo and what we'd be in for. I don't think any of us really believed it until they told us for sure that's where we were going. She broadcast every night and we listened on the short wave radio aboard the ship.

"She could speak perfect English and she let us know what we were in for. She said, 'Our boys are just sitting back, reading mail from home, waiting for you. All that bombing you have done hasn't hurt them at all. They are underground, just waiting.'

"We didn't believe her, but she was tellin' the truth. The Japanese had tunnels all over the island. They were prepared for a fight…a long one. We all thought she was spouting propaganda, scare tactics. She also said she knew 'where the 5th Marine Division was going' and told us we 'had a one way ticket.' She was right about that too; a lot of those fellas I was with never made it off Iwo."

Mom's bedroom door cracked open a few inches.

"Are y'all by yourselves?" she asked.

"Yes Mom, it's just us."

Her door opened, and she came out carrying her water glass and made a beeline for her chair in front of the fireplace. I could see why she asked if we were alone - she had her flannel gown on and her hair was a sight. Mom doesn't like to be seen by people, if she's not *fixed up*, as she calls it. Apparently, Dad and I didn't count as people.

He waited until Mom settled into her recliner before he continued.

"When we were headed to Iwo, I learned later, the brass thought it would take just a few days to take the island. They were going on how long it took to take Saipan and the other islands...but Iwo would be different. Of course, we didn't know that then.

"Also, afterwards I found out the plan was for the Marines to go in and cut the island in half the first day, then the second day part of them go north and part of them to Suribaci, then the third day, wipe up and clean up. But it didn't happen that way.

"When we got on that troop ship, you just kind of got caught up in another world. Some of the guys had been in combat before, but not us Seabees. The Seabees were new. Everyone was in a pretty somber mood.

"We were in a huge convoy traveling to Iwo. There were carriers, subs and, of course, numerous troop ships and battleships. Over eight hundred ships took part in the bombardment of Iwo Jima and that's not countin' all the landing crafts. It was a sight to see...all those ships, guns, and men.

"We kept movin' closer and closer to the island. On the mornin' of the nineteenth I couldn't sleep, so I went up on deck about 2:30 a.m. I was on the ship's main deck watching the barrage. I thought to myself, very few people get to see that much firepower at once. Those 14 and 16 inch guns on the battleships are so big that a man can slide down the barrel...and they were firin' three of them at once. It was quite a show.

"For the first wave that was goin' in, they started serving breakfast at about 2:00 a.m., so the guys would have enough time to get their gear together and get aboard the landing crafts. The skipper made a speech before the invasion began. You could hear it all over the ship through the loudspeakers. I was still on deck listening.

"I'll never forget his final words: 'Good hunting.'

"Lookin' back, I realize those words would haunt me…you can't hunt someone you can't see. We were going to fight someone we couldn't see. Just like Tokyo Rose had said, they were underground, just waiting.

"All of the battleships, cruisers, and destroyers pounded the island until after daylight. The intense bombing had blown all the Japanese big guns and anti-aircraft guns on the surface out of existence or rendered them useless. But they still had those guns they could run out of their underground holes and fire.

"When it got daylight, a squad of B-29s came over and they strowed bombs from one end of the island to the other. The first Marines loaded on the landing crafts at 7:00 a.m. and just before they hit the beach, some fighter planes came off of one of the aircraft carriers and strafed the beach from one end to the other where the Marines would be landing. A small ship also laid down a smoke screen to try and hide the Marines when they went in, so the Japanese couldn't see them coming.

"I watched as the first wave of Marines loaded into the Higgins boats and headed towards the beach. I could hear some of them singing:

Oh, what a beautiful morning;
Oh what a hell of a day.
I've got a horrible feeling…
Something is coming my way.

"I guess it was their way of keeping their spirits up, but it sounded strange echoing over the waves, with all that blastin' going on."

The image of those young Marines setting out across that churning deep through the thick smoke and deafening volley of the battleships with the buzzing overhead of bombers made me sad - knowing how few of them would return.

"The first wave got in pretty good," Dad said, "but as the second wave loaded and headed in, the Japanese came out of their holes and began firin' on the beach, and by the third wave…it was terrible…wave after wave of our men were just slaughtered. I couldn't see anything from where I was, but I could hear it. We had to stay out of range of their guns. The earlier bombings had blown all their big shore guns off or out of commission, but the Japanese still had these big guns they could run out of their caves and shoot at us. They were on tracks and they would run them out, shoot and pull them back in.

"I saw them bringin' the wounded back and knew it was bad. They couldn't bring the dead back…they just had to leave them there." He took a deep breath.

"They couldn't get any of their equipment off the beach. The heavy guns they needed against the Japanese were unloaded and immediately sank into that black sand. It wasn't like regular sand. It was more like volcanic ash. That was when they called in C Company of the 31st battalion, my company, to do something about the beach and all that sunken machinery. We were the heavy equipment boys; that's why they called us in to try and clear the beach.

"This was something they hadn't foreseen. Frogmen were supposed to have done a reconnaissance on the landing area and report any potential problems. Apparently they didn't know that the light volcanic ash wouldn't hold heavy weights. It was chaos. We were called in during the heaviest fire of the invasion.

"The third mornin' we were loading to make the landing. Even gettin' into the Higgins boats was a dangerous proposition. We had to go over the sides of the ship on rope ladders and get into the small landing craft. Doesn't sound too bad until you consider the huge swells in the ocean, and with every one of them you are tossed against the side of the ship. Several men drowned or were crushed to death, just transferring into the boats. We had our packs, our guns, all our gear on, so it was awkward trying to climb over the side and down those ropes.

"One of our guys lost his grip and started falling, and a couple of guys managed to grab his arms and hang onto him. He lost some of his gear in the water…it was an eerie sight to see his coat and other stuff pulled into those huge ship propellers.

"There were about thirty of us in the Higgins boat. The sides were about four or five feet high." He measured the depth with his hand from the floor as he talked. "The coxswain told us to 'Get your heads down!' but we couldn't get down very far. We were already standing in water and it was February and pretty cool. When we finally headed in, the Beach Commander on one of the destroyers yelled at us through his bullhorn. 'Hold up! The beach is too hot!' So, we circled and waited…our boat was damaged from earlier runs and we began to take on more water.

"Finally, the coxswain maneuvered our boat around and pulled up beside the destroyer and yelled through his bullhorn to the Commander, 'We're either going in or we're going down!' The Commander yelled back, 'Hit it and get the hell out!'

"When you're in the middle of something like that, you don't have time to size up the whole operation; you only know what is happening in your boat. When we went in, we all had an assigned place to land. We landed on the upper end of Red Beach, next to Green Beach. We were close enough to be under fire from Suribaci and the sniper fire. The shells from the Japanese big guns they had on tracks could reach all up and down the beach. A lot of people think that if you weren't on the 'front lines' on Iwo, that you weren't in danger...but that wasn't true on Iwo Jima – the Japanese could hit you anywhere on the island from where they were. It didn't matter if you were on the so-called front line, the beach, the interior...on Iwo there were no safe zones." He sighed.

"Before we hit the beach, we could hear shells hittin' the water all around us...ploomp! Ploomp! When we finally did make it to the beach, the water ran to the back of the Higgins boat as we went up onto the edge of the beach and it sank. The ramp wouldn't let down because it had been damaged in earlier runs. We had to climb up and over the sides of the Higgins boat into the water and try to run up the beach.

"The whole time we were trying to get over the sides you could hear bullets poppin' all around us. It was pretty risky business going over the sides. Hitting the beach, we scrambled up the slope; the sand was jumpin' on every side from the bullets. We kept running until we found a bomb crater to get in. Running up the beach was like trying to walk in a tub of shelled peas...every step you took was a struggle. That volcanic ash would just fill in around your feet and kind of suck you down, like dry quicksand. The coxswains were stuck...they had no boat and had to run like the rest of us to try and find shelter from the heavy fire.

"Trying to survive was all we could do. It was chaotic. No one was in charge for a while. When you're in a situation like that, you become a different person. You go into a place where you can deal with what is happening right then. You have to block out the horror of the moment for a while or you can't think. It's running on instinct, I guess you'd call it. It's the only way you can get through something like that. The beach was covered with dead men. There were so many. It was like that up and down the beach...but you can't think about that when you are running for your life."

Dad shook his head. "It's hard to imagine what it is like, unless you've seen it with your own eyes." His voice trailed off while he stared into the flames of the fireplace. In the silence that followed, broken only by the crackling of the logs on the fire, I tried to picture the scenes he had described and how terrifying it must have been to climb over the sides of that Higgins boat into the water, praying you didn't get shot, and try to run up the beach with bullets hitting all around. I remembered in *Saving Private Ryan* there was a scene much like he described, and as I thought about it, a lump formed in my throat. The good Lord must have certainly had a plan for Dad because it is a miracle he made it off that beach alive when so many men didn't.

Dad roused out of his thoughts. "You know," he said, "the brass had told everyone this would be a three-day battle, but that didn't happen. We were still fightin' the Japanese on Iwo months later - not many people know that. I was on that island for eight months, until Japan surrendered."

He looked over at me. "Didn't go like they'd planned, did it?"

I shook my head. "Sure didn't. What a different kind of battle it turned out to be."

"Anyway, back to the beach," He said as he turned back to gaze into the fire.

"When we finally made it up the beach a ways and found a mortar crater, we dove in. We had to try and protect ourselves from the almost constant fire, so we grabbed what we could to try and make our foxhole more secure. We used some of that useless steel matting to put in the sides of the hole to try to keep from gettin' buried by that volcanic ash. It was like being in a vat of dried corn or peas; every move we made, it fell in on us and we couldn't keep it out. When we were trying to fix our foxhole, we would have given a month's wages for some sand bags, but none were available. We did use some of the bent and useless matting for a makeshift top on our hole, but we had to use scattered bedrolls and blankets that we could reach nearby to line the matting and try to keep the sides from cavin' in. There were plenty of blankets…dead Marines were all around us. It would be days before we could get them rowed up and covered."

He paused as one of those looks went across his face. It was that same inscrutable look I have seen before when he was approaching a subject he was reluctant to talk about. I knew it was best to just give him some time, so I turned to stare into the fire and consider what he had already shared. I don't know how he processes those terrible memories or if he just shuts them away in a private place and locks the door. So many of the veteran's family members I talk to say their loved one never talks about their experiences. Hearing some of the things Dad went through, I can understand why.

Dad cleared his throat and asked Mama if she was warm enough. She said she was, but needed a glass of water. As I walked to the kitchen to get it for her, I could hear them talking in low soft tones. It seemed to help him move past whatever had caused him to stop talking because when I came back into the Big Room with Mom's water, he asked me to bring him some peanuts, his favorite snack. I went back into the kitchen to get the big jar he always kept on the counter, but it wasn't there. Then I remembered he sometimes carries it into his sunroom. Sure enough, there were the peanuts by his chair.

"Here you go," I said as I walked back in, holding the jar aloft. Mom poured some into a bowl she keeps by her chair for crackers, then I handed the jar to Dad and sat back down. When everyone was settled in, he continued his story.

"When we were coming in on the beach, some of our guys had to unload our heavy equipment. One of our dozer operators, Alan Lockett, said he felt something on his side and his gas mask fell off. A bullet had cut through the strap, but it had missed him.

"Another operator, a guy from West Texas, Ben Massey, wasn't so lucky; he was shot through the neck. It didn't kill him, but that was the end of his time on Iwo. Those bulldozer operators were sittin' out there in the open with no protection…sittin' ducks for the Japanese snipers up there on Suribaci.

"We had to have the bulldozers to cut a path up the beach." Dad motioned with his hand, making an upward sweep to his right. "They had to go at an angle to get up and over those humps on the beach to level ground. That was the only way we could get the equipment off the beach and to the guys who needed it. The Marines had tried going straight up the beach and had just bogged down in the soft ashy sand. There were hard clay banks that kept the tanks from going further in the island and there were just certain spots you could maneuver up and over those banks. The Japs had placed land mines in those areas, and the few tanks that did make it off the beach didn't make it far. In one spot, I remember, there were four tanks just blown inside out from hittin' those mines…those men never had a chance - they didn't know what hit them." He shook his head and stared into the fire.

"The 133rd CB battalion had a truck full of men that ran over a land mine later…they lost a lot of guys." He had a pained expression and paused before he continued.

"That volcanic stuff on Iwo was something they hadn't encountered before. The Marines had laid some steel matting down, but when they tried to take the heavy vehicles up at that steep angle, the steel matting just wadded up underneath them. Those steel mats work wonders in a marsh, but up that steep volcanic ash, they just wadded up." He shook his head again. "It was a sight."

We all got quiet for a minute.

"Remind me, what kind of equipment were they trying to get onto the island?" I finally asked just to break the silence.

Dad looked a little startled when I spoke, like I had roused him out of some sort of private recollection, but he continued his story without mentioning his thoughts.

"The Marines had to have their big guns if they wanted to get to the Japanese barricaded in those caves. Their weapons carriers were all-wheel drive vehicles and they had their big guns on them and their ammunition carriers, but all of them buried up in that volcanic ash. Even walkin', sometimes you'd bury up to your knees. It was slow goin'. I think that's one reason so many got killed on the beach…you just couldn't get off there fast enough.

"My buddy Clifford Paxton went in on the first day. It wasn't until I saw him several weeks later that I found out what had happened to him. Here's what he told me:

"Clifford said him and DeRamus went ashore that first day in an LCVP. It was loaded with all kinds of ammo, diesel, and gasoline. Clifford said they unloaded around 4:00 p.m., but when they started to raise the ramp to leave and go back to the ship, only the right side would go up.

"Clifford told me that him, DeRamus, the coxswain, and an ensign worked for about twenty minutes trying to get the left side up. 'Finally,' Clifford said, 'the beach master started yelling at them to get that so and so off the beach. But as they started backing off to abandon the boat and let it sink, all of a sudden the ramp came up.'

"I remember Clifford said, 'God must have been awful busy that day, but He was there for us.' Then, when they finally got away from the beach and went out to where their ship was supposed to be, it was gone because the APAs all leave and go out away from the cruisers and battleships for fear of an air raid."

"What's an APA?" I interrupted.

"An APA is what the Navy called its attack transports. They carried men like a troop ship. A lot of the APA's were converted from troop ships to use as attack transports during WW II," Dad answered.

"And poor old Clifford and his bunch found themselves all alone out there just circling. Sure enough, that night there was an air raid. Thank goodness they didn't get hit. They also didn't have any food or water because they expected to be back on their ship by night. So they pulled up alongside a battleship and asked them for some rations.

"Clifford said they threw down some water, bread, and some Spam, that potted meat they had in a can. They spent the rest of the night dodging the big boys, circling Iwo. He said it was one of the longest nights he could remember."

"Poor Clifford," I said and shook my head. "It's a miracle they made it through the night."

Dad nodded. "There were a lot of things that happened those first few days that I doubt any of us will forget."

"What happened after you came on shore?" I asked. "What did you see?"

He leaned back in his chair and closed his eyes, as if trying to recapture the memory from the mists of the past. Mom and I waited in silence.

Finally he began to talk again.

"One of the first things I remember seeing after getting' on the beach was this camera on a tripod standing there all by itself…apparently the reporter or whoever it belonged to had already gotten hit." He shook his head. "You've got to admire those guys; they are right there, life or death, in the thick of things.

"I found out later, there was this famous reporter that got killed on Iwo Jima. I don't know if that was his camera and tripod, but it was on Red Beach where I landed. It was a big camera, a professional one, just like Rosenthals' who took the picture of the flag-raising. Those reporters were right there alongside our guys, doing their job. Some people like living on the edge I guess. I was living on the edge too…but I didn't volunteer for it.

"We were under intense fire day and night those first few days. We were just fighting to stay alive. Even going to the bathroom was a life-threatening challenge. We had no building or anything; they dug a small trench and put up a rail about two feet high and you hung over the rail. We had to have something because of sanitation concerns, but you can bet we didn't linger on the rail. You were under fire all the time…you just did the best you could. It was kinda like the children of Israel traveling in the wilderness, except they weren't getting shot at." Dad almost smiled.

"It wasn't funny at the time though. No one, except those of us on the island, would know the horror of those first few days and weeks. Folks back home had the idea that when the flag went up on Mt. Suribaci that the battle was won, but they were wrong…it was just getting started. I guess they had to give something positive for those on the home front, because at that point, things weren't lookin' too good for our side.

"We were under strict orders to remain in our foxholes, especially at night, because they would shoot anything that moved. There were only supposed to be two men to a foxhole, but we had three…Clarence Peterson, myself, and a demolition fellow by the name of George Holliday - he was from a little town in Washington State. Later on, it was just me and Pete, but for those first few days, we managed with three.

"We were about 200 yards off shore and dug in when we began to receive sniper shots from above our foxholes. We couldn't tell exactly where it was comin' from because he'd only take four or five fast shots and that was it.

"On the third evening it happened, three of the guys were waiting up the hill for him to take his pot shots…he was pretty clever, using the setting sun as his shield. As soon as he slid back down into his hole, our guys rushed the hole and unloaded three grenades right after him…we didn't have any more trouble from him."

We sat there quietly for a minute and then Dad yawned and closed his eyes. A few seconds later, so did Mom. That was their signal that it was time to start thinking about lights out.

I checked my watch and realized I had to go home and feed my animals. "Dad," I said softly. "I have to go on home, but could we talk some more tomorrow?"

He opened his eyes and looked over at me. "You mean you're gonna leave me in that foxhole all night?" He raised his eyebrows and leaned towards me, then broke out in a grin. "You go do what you've got to do; I'm not going anywhere," he said. "Well…maybe down to the home place to check the cows tomorrow, but that's all."

"Ok," I grinned back. "I'll holler at you tomorrow."

Calling out one last *Love you* to Mom and Dad, I quietly shut the door behind me, but not before turning the lock. Guess I'm like Mom - hedging my bets against all possible scenarios. I want to make sure they are safe and sound…I will sleep better if I do.

PHOTOS

Image 1:
Left to right back row: Robartus C. McAnally (Stuart's Dad), Henry McAnally (Dad's oldest brother who was already grown when they moved to Jacksonville in 1927), Edna Young McAnally (Stuart's Mother or Ma, as we all called her) holding Oscar, standing next to Ma is Eva, then Josie, Edith "Babe", Ura "Sis" Buffalo (Stuart's oldest sister) and her husband Leo Buffalo.

Left to right on the front: Stuart (Dad), Monroe, Floyd (the taller boy), Delbert (the small boy in front of Floyd) and the small girl in front is Sis and Leo Buffalo's only child Agnes.

Photo taken circa 1934

Image 2:
Dad with a sawmill crew before the war. He is sitting on the right, closest to the log truck. He was eighteen years old when he drove the log truck for sawmill owner, Coy Grimes.

⇑*Image 3:*
Dad in his Seabee uniform, taken in California – 1944

⇐I*mage 42:*

Dad in his Marine work clothes while serving in Occupied Japan.

Image 4:
31st Shore Patrol unit – Newland (4th from left), George Elm (5th from left), Clarence "Pete" Peterson (2nd from right), Stuart "Tex" McAnally (4th from right) and Cecil "Coon" Watkins (5th from right). The 31st patrol was made up of men from all the different companies and units, so we were only able to positively identify

PATROL

these men. George Elm was a husband and father of three who became ill while serving on Iwo. They sent him home from Iwo, but he died a short time after his arrival in the states, apparently from some sort of cancer. Peterson, McAnally, and Watkins all returned home eventually and married their sweethearts.

Image 5:
Dad and Cecil "Coon" Watkins on liberty in Hollywood – 1944.

Image 6:
Dad and his buddies out on the town in California – 1944.

Bottom row left to right: J.P. McEwen (Louisville, Kentucky), George Elm (Boston, Massachusetts), Cecil "Coon" Watkins (Lake Charles, Louisiana)

Top row left to right: Stuart McAnally (Jacksonville, Texas), Clarence Peterson (Chicago, Illinois), and Ira S. Wall (Shamrock, Texas)

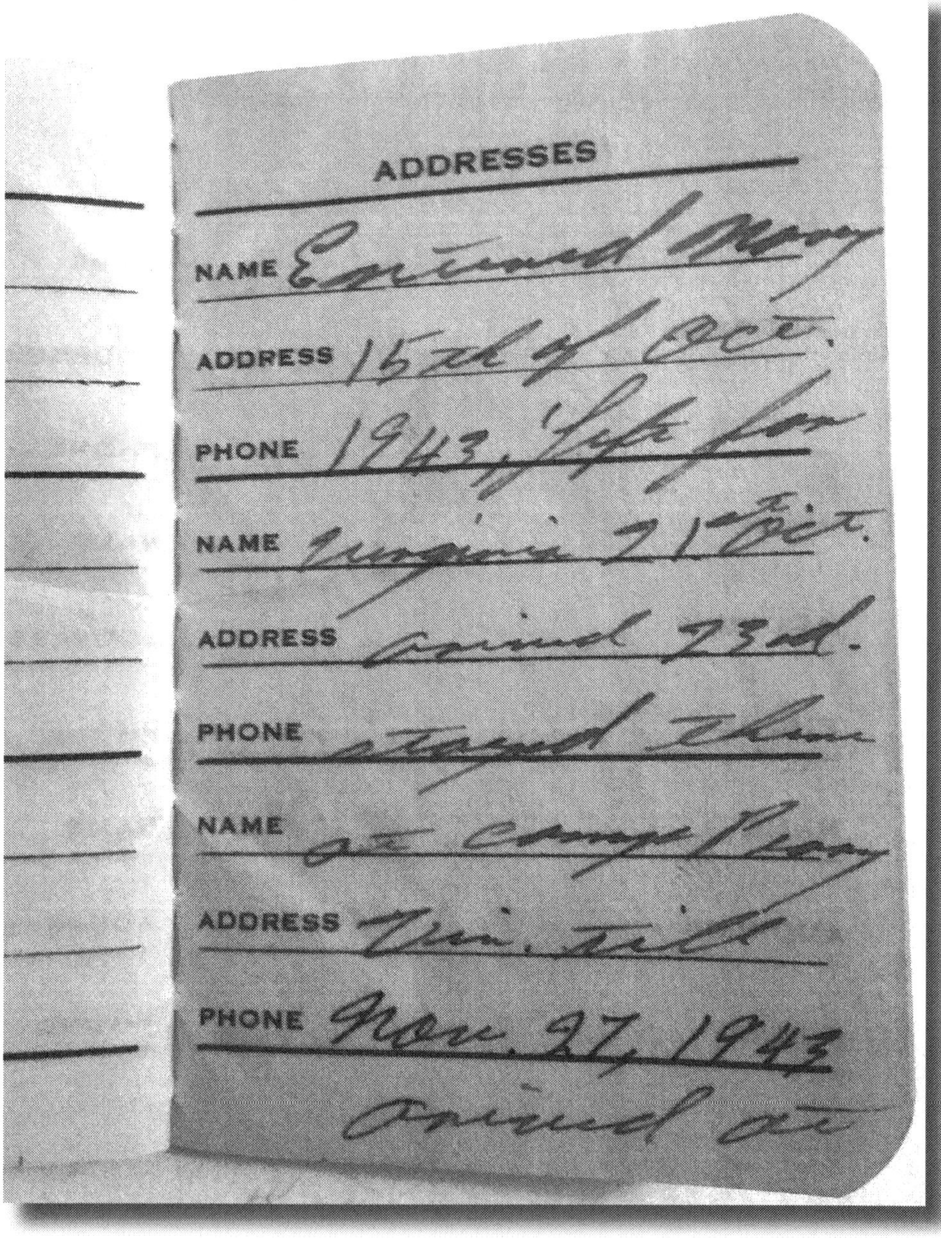

Image 7:
Pages from the journal Dad kept during the war.

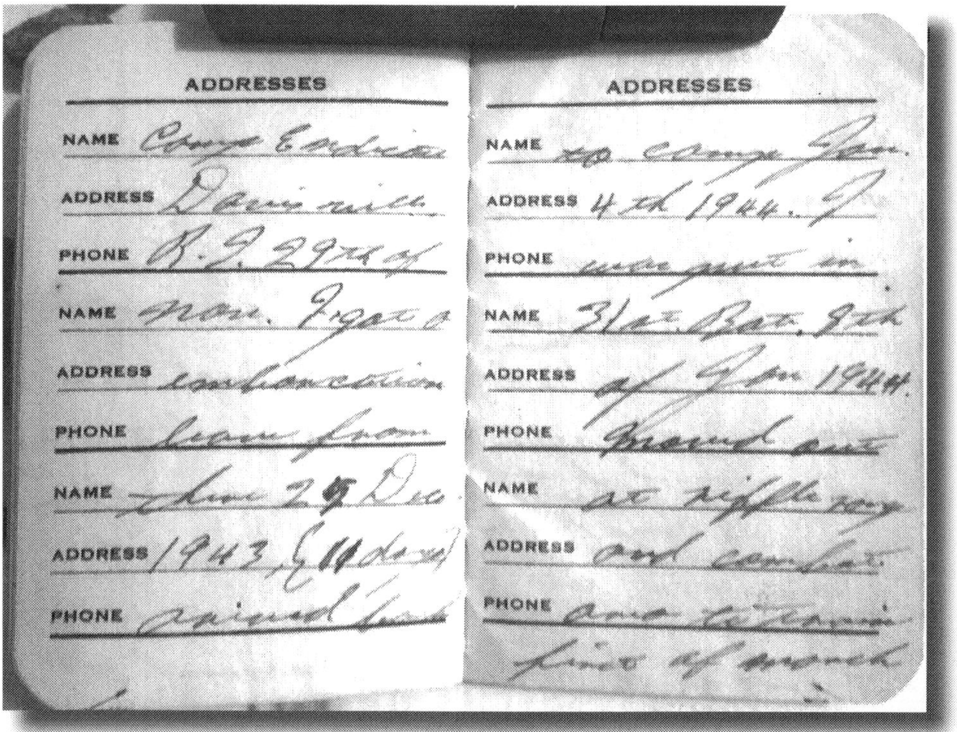

Image 8 & 9:
Pages from the journal Dad kept during the war.

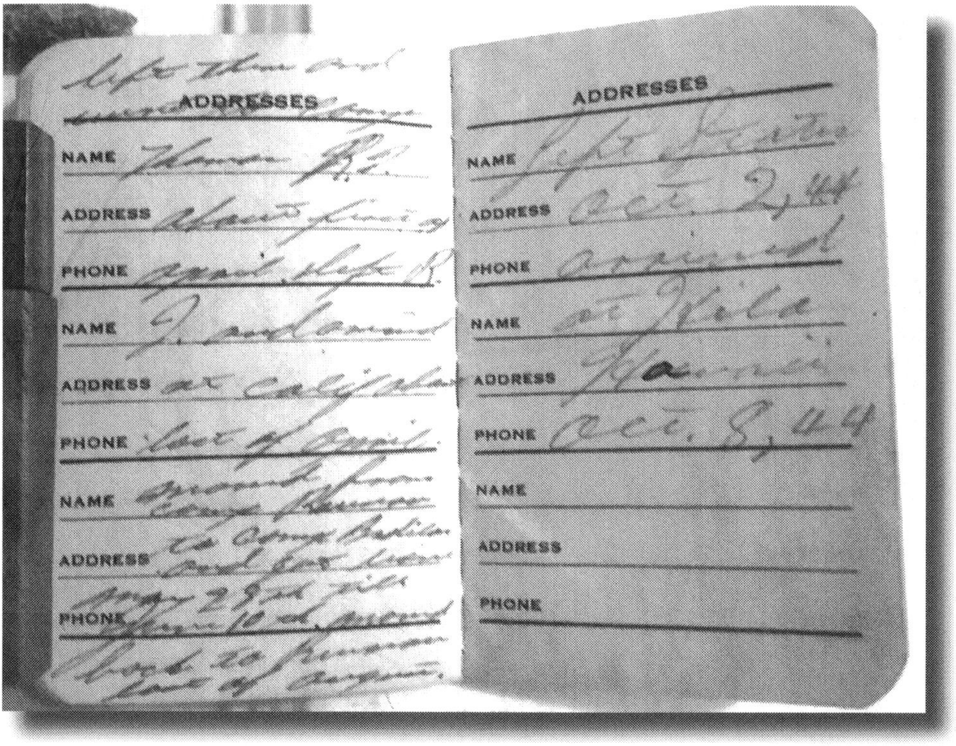

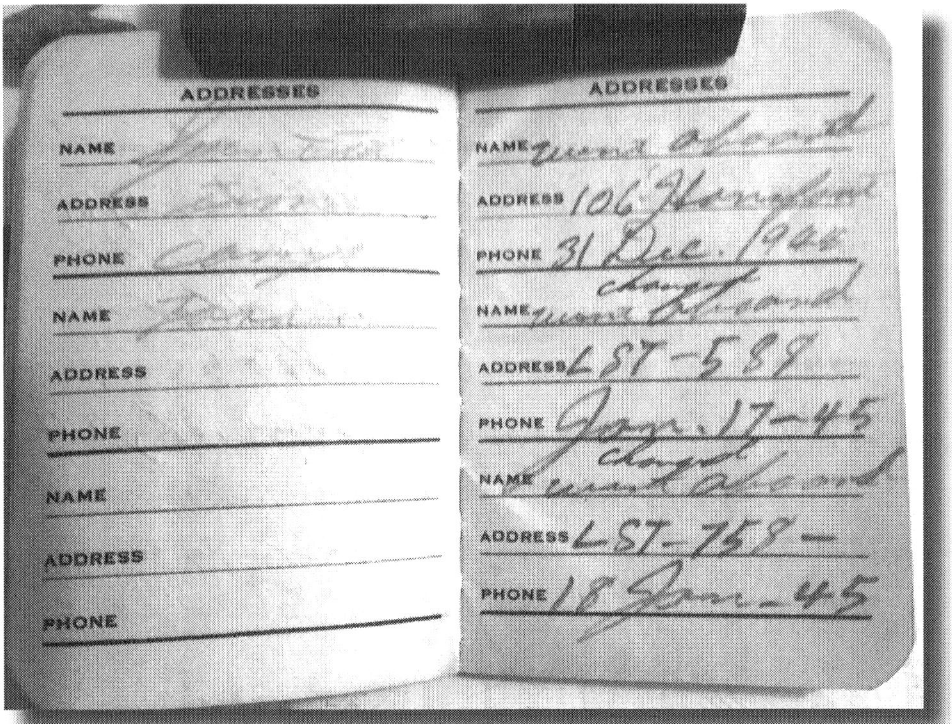

Image 10 & 11:
Pages from the journal Dad kept during the war.

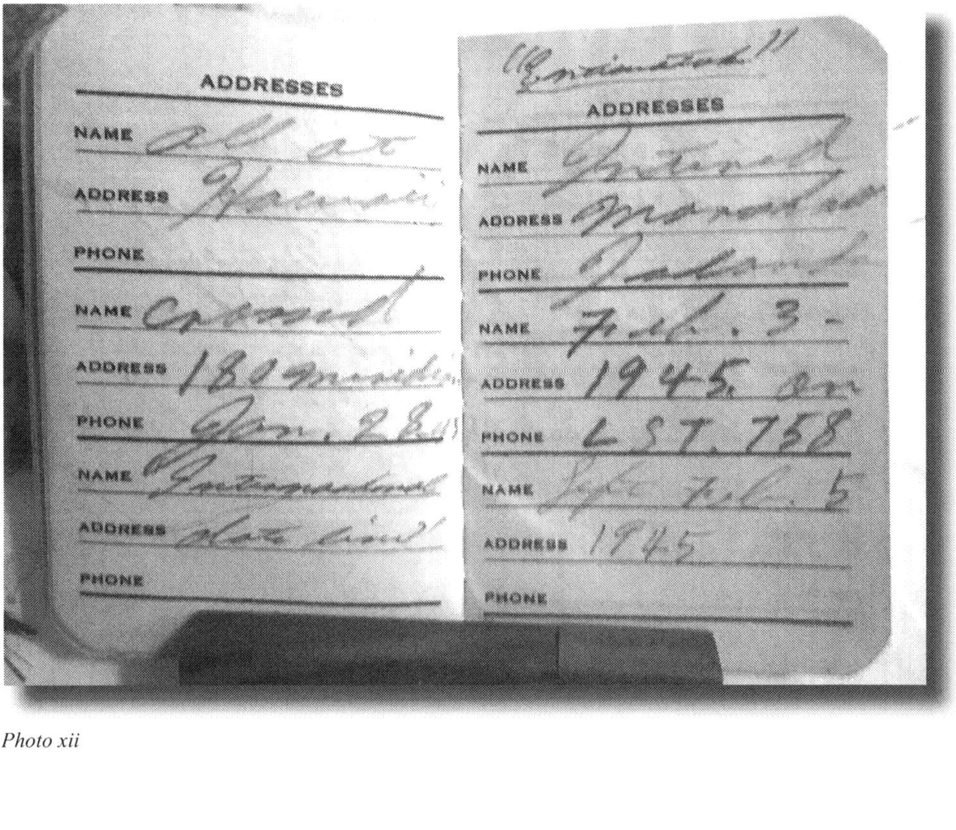

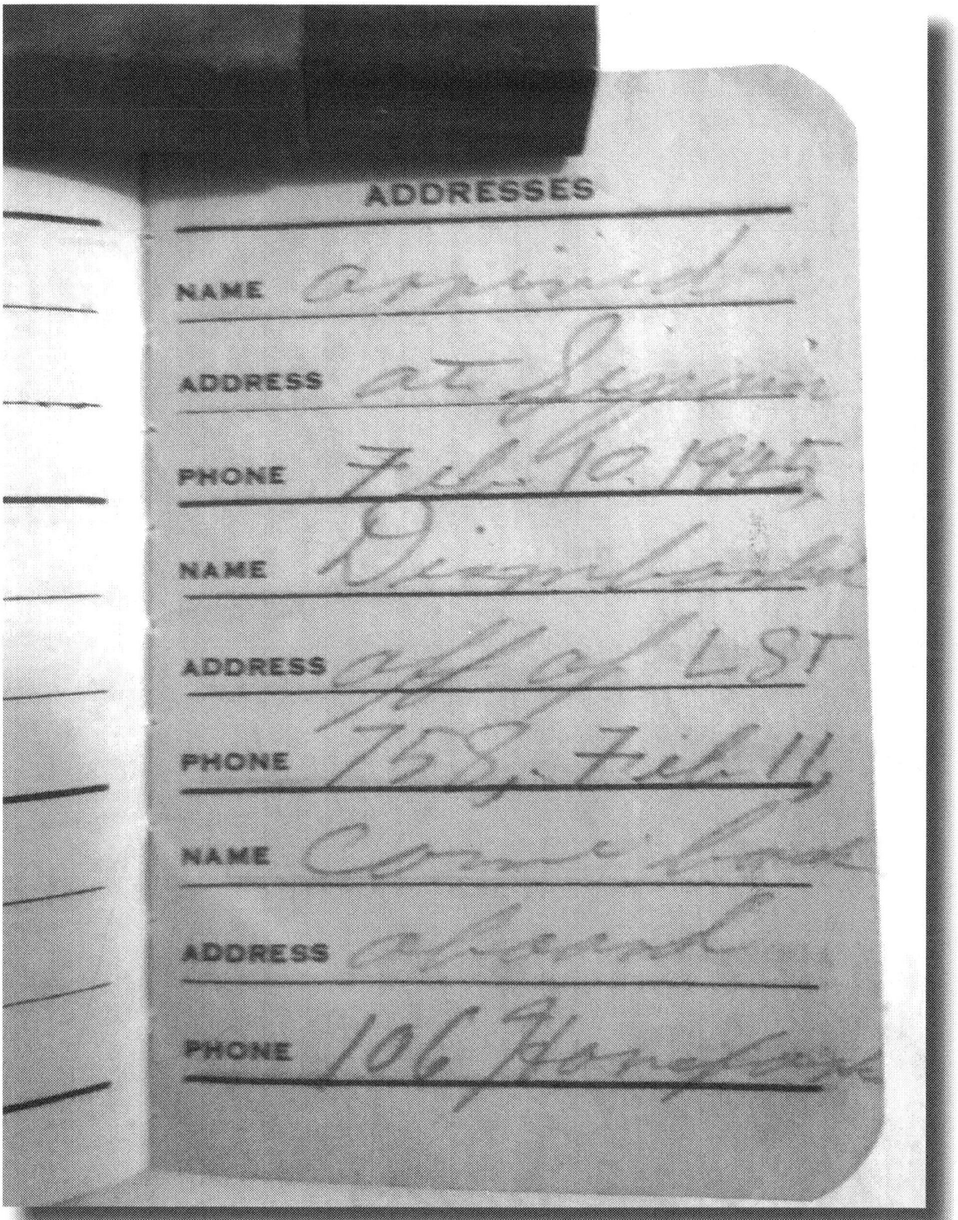

Image 12:
Pages from the journal Dad kept during the war.

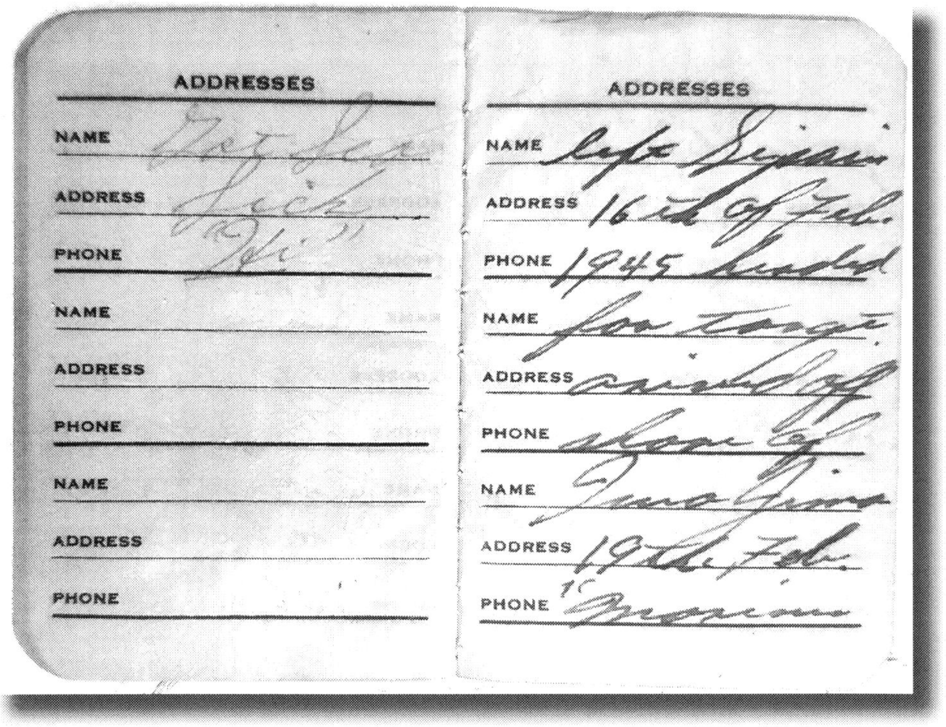

Image 14:
Pages from the journal Dad kept during the war.

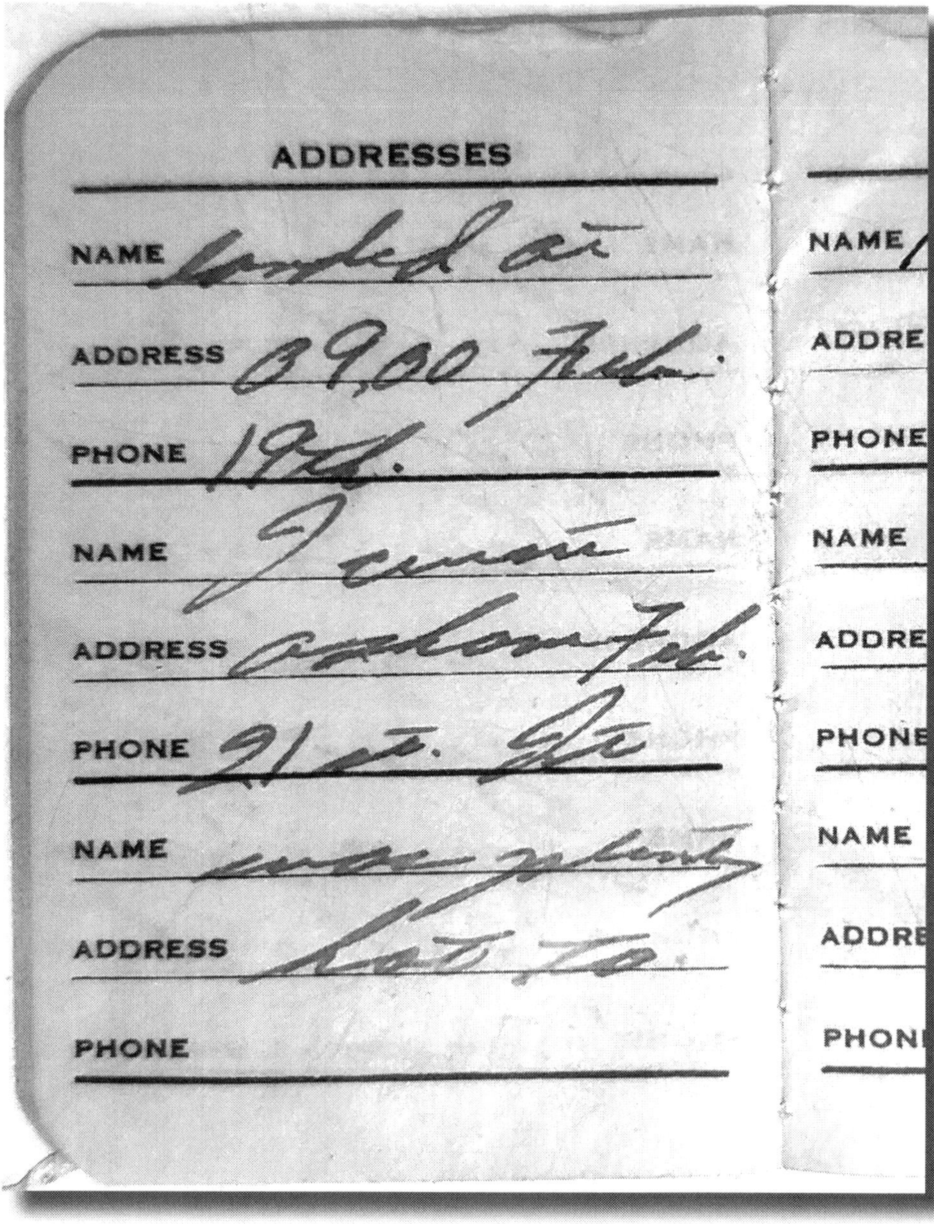

Image 14:
Pages from the journal Dad kept during the war.

Image 15:

The crew Dad worked with, led by Ensign Pierce, heading for Airstrip No. 1, where they got shot at by Japanese sniper and had to beat a hasty retreat. This is when Steed was shot in the leg. (Taken from the 31st CB book)

Reveille did not need to sound on D-Day morning. Men were up early for their breakfast of steak or ham and eggs, a concession to the misery that lay ahead. The approaching job began to take on reality when we heard the distant blasts of ship's guns, and looked out to see the inky blackness of the early hours punctured with balls of red and yellow billowing fire. As dawn stole in unnoticed amid the noise, she let us see the outline of the island ahead—Suribachi on our left. At a safe distance from the counter-fire of shore installations, coveys of small boats were lowered over the side, and went into their merry-go-round act near the mother ship, as they waited for their loads.

The designated time approached; invasion parties were called up from their compartments below, and lined up ready to climb the ladders and cargo nets.

Above: 31*st'ers start toward No. 1 airstrip to clear it of shrapnel* (upper left) *while it is still under fire.*

Image 20:
C Company – Platoon Four (Dad's group)

PLATOON FOUR: ***Abbe.*** **Row 1:** ***F. H. Edwards, German, Gentry, Bullock, Arledge, DeBerry, J. M. Brown, L. A. Brown.*** **Row 2:** ***Hogg, Grosjean, Pester, Ulrich, Holliday, Blosser, L. E. Taylor, Guin, S. C. Clark, Garrett.*** **Row 3:** ***Spaulding, Viehmann, C. W. Petersen, R. C. Garner, Stull, McFalls, Wightman, McAnally, Lockett, Wilkinson, W. C. Wood.***

Image 21:
Names in Platoon Four

⇑*Image 16:*
Dad is in this group clearing the airstrip. (Taken from the 31st CB book)

Above: 31*st'ers start* to
***it of shrapnel* (upper lef**

Photo xviii

⇓Image 17:
Original caption for photo above

No. 1 airstrip to clear
hile it is still under fire.

Photo xix

Dear Sister

Image 18:
The first letter Dad wrote home from Iwo Jima. It was to his older sister Edith "Babe" Harper. He was still in his foxhole and would be for almost another month

Photo xx

MAR
20
1945
NAVY
MAIL
UNITED STATES OF AMERICA
VIA AIR MAIL
306

Photo xxi

Hill No. 362 was the high point of the broad part of the island. Located at the south end of what is now the third airstrip, it was a deadly Marine objective. When taken, it served as a radar site. Its jutting promontory housed a great cave system which is mapped on the opposite page. It was here that "Whitey" Kneer gave up his life. Found in a cave under which this dozer is working, he was buried in the cemetery of the Fourth Marines.

Image 19:
Part of the extensive cave system the Japanese had on Iwo Jima. This is where Whitney Knerr's body was found. (Taken from the 31st CB book)

Iwo Flag-Raising Scene Reproduced by Sculptor W. T. Rich on Side of a Cliff.

Image 22:
W.T. "Art" Rich and his now famous carving of the flag-raising on Iwo Jima; it can still be seen today.

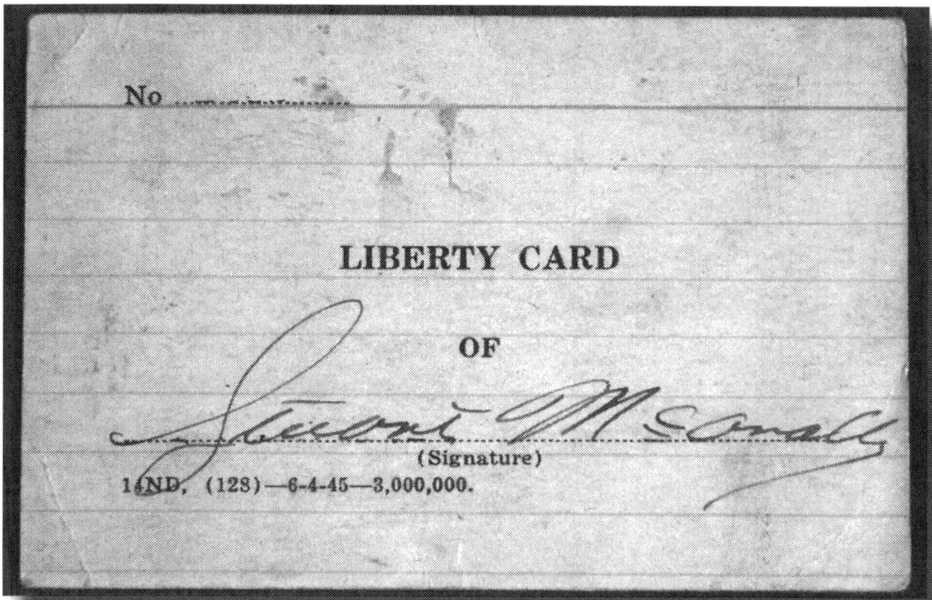
No ....
LIBERTY CARD
OF
(Signature)
14ND, (128)—6-4-45—3,000,000.

Image 23:
One of Dad's Liberty cards.

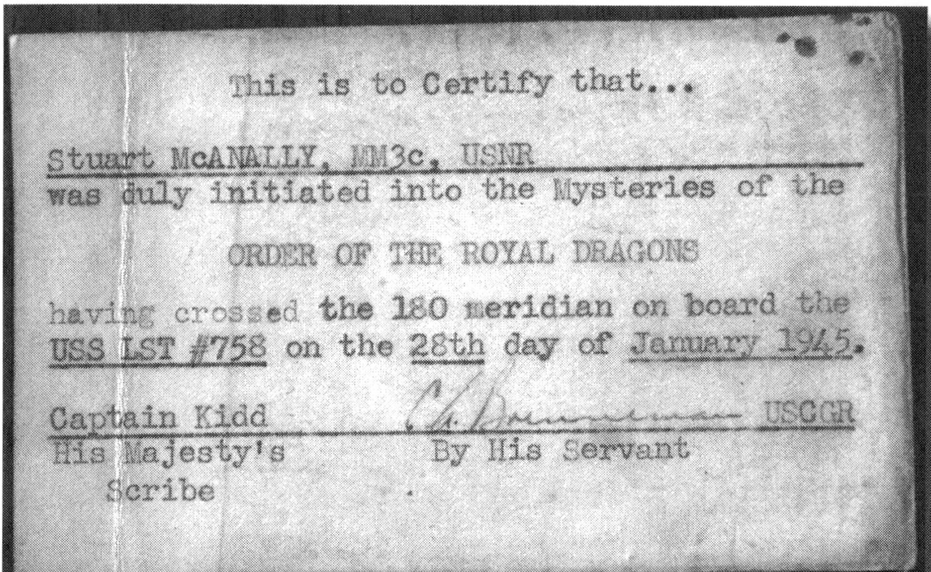
This is to Certify that...

Stuart McANALLY, MM3c, USNR
was duly initiated into the Mysteries of the

ORDER OF THE ROYAL DRAGONS

having crossed the 180 meridian on board the
USS LST #758 on the 28th day of January 1945.

Captain Kidd USCGR
His Majesty's By His Servant
Scribe

Image 24:
The Order of the Royal Dragons card he received after crossing the International Date Line.

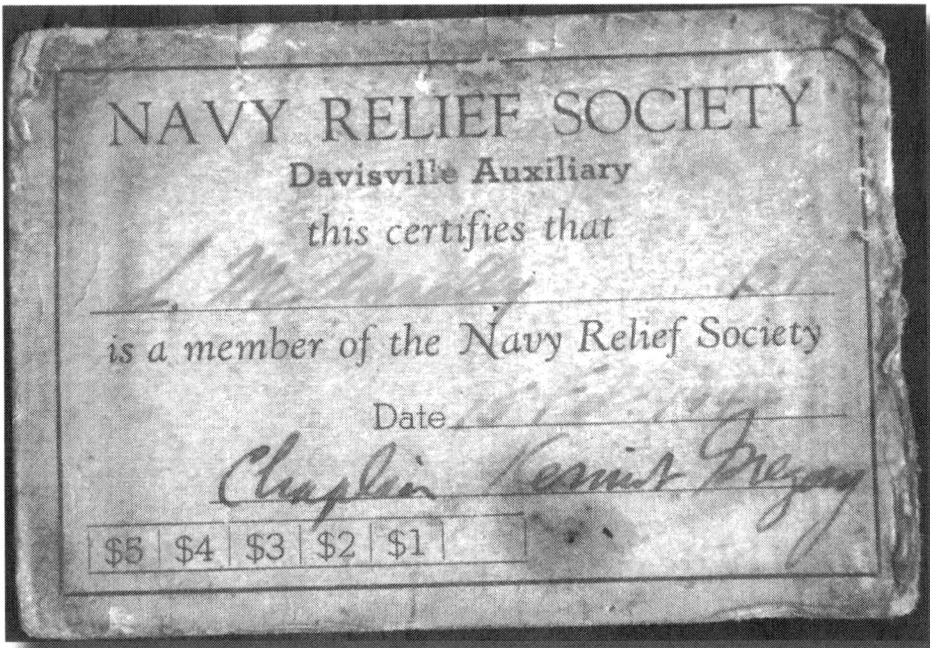
NAVY RELIEF SOCIETY

Davisville Auxiliary

this certifies that

[illegible]

is a member of the Navy Relief Society

Date [illegible]

Chaplain [illegible] Gregory

$5	$4	$3	$2	$1

Image 25:
Naval Society Relief card he carried while at Camp Endicott, Rhode Island, signed by Chaplain Gregory.

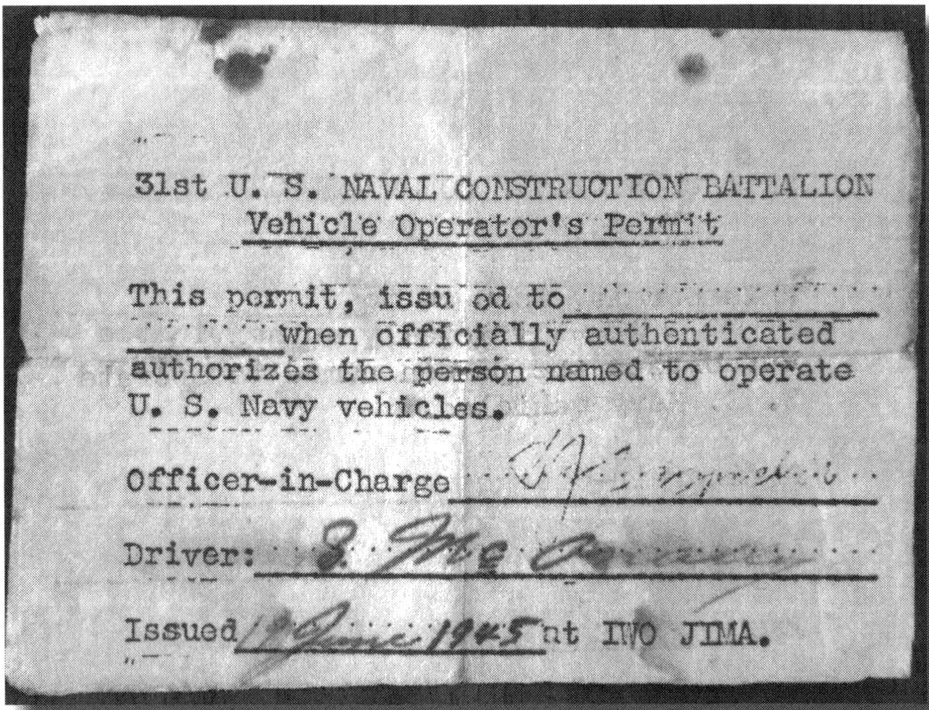
31st U. S. NAVAL CONSTRUCTION BATTALION

Vehicle Operator's Permit

This permit, issued to [illegible] when officially authenticated authorizes the person named to operate U. S. Navy vehicles.

Officer-in-Charge [illegible]

Driver: [illegible]

Issued [illegible] June 1945 at IWO JIMA.

Image 26:
Dad's driver's permit issued to him while on Iwo Jima.

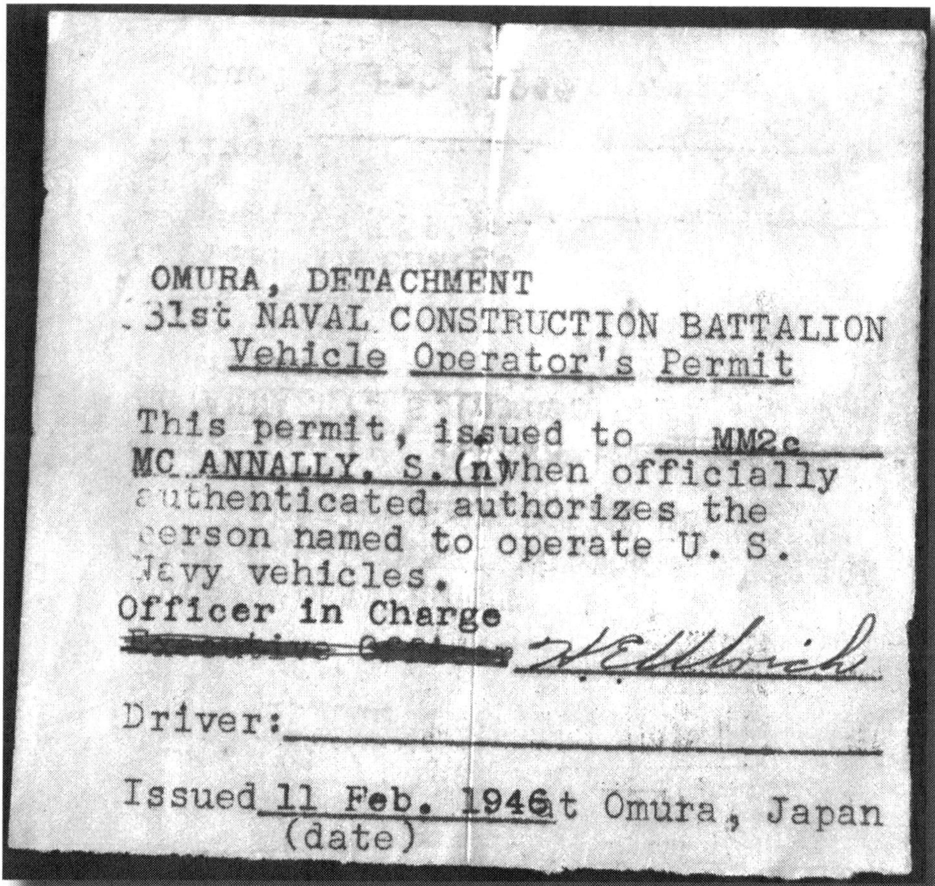

OMURA, DETACHMENT
31st NAVAL CONSTRUCTION BATTALION
<u>Vehicle Operator's Permit</u>

This permit, issued to MM2c MC ANNALLY, S.(n) when officially authenticated authorizes the person named to operate U. S. Navy vehicles.

Officer in Charge
~~[illegible]~~

Driver: ______

Issued 11 Feb. 1946 at Omura, Japan
(date)

I*mage 27:*
Dad's driver's permit issued to him while in Omura, Japan.

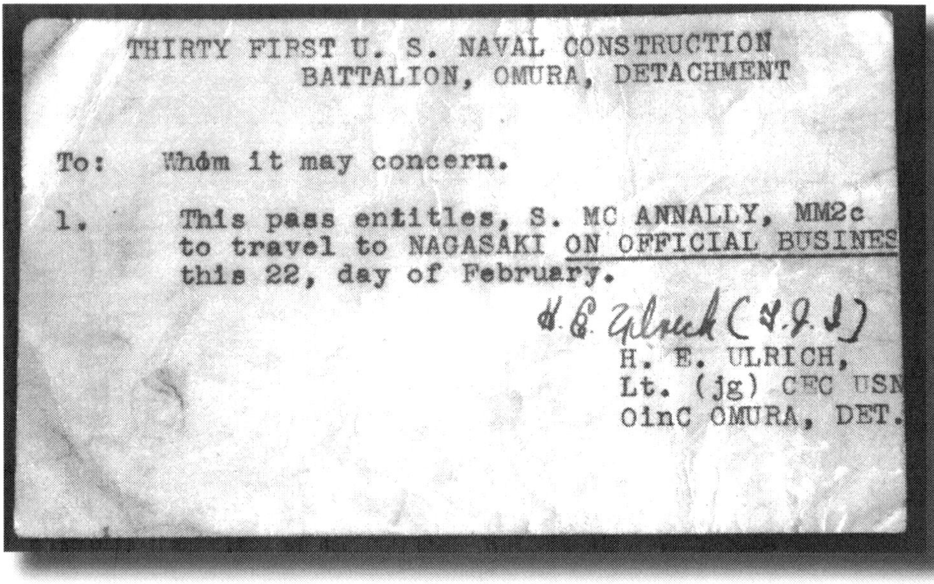
THIRTY FIRST U. S. NAVAL CONSTRUCTION
BATTALION, OMURA, DETACHMENT

To: Whom it may concern.

1. This pass entitles, S. MC ANNALLY, MM2c to travel to NAGASAKI ON OFFICIAL BUSINES this 22, day of February.

H. E. Ulrich (?)
H. E. ULRICH,
Lt. (jg) CEC USN
OinC OMURA, DET.

Image 28:
Pass issued to Dad when he went to Nagasaki to pick up equipment.

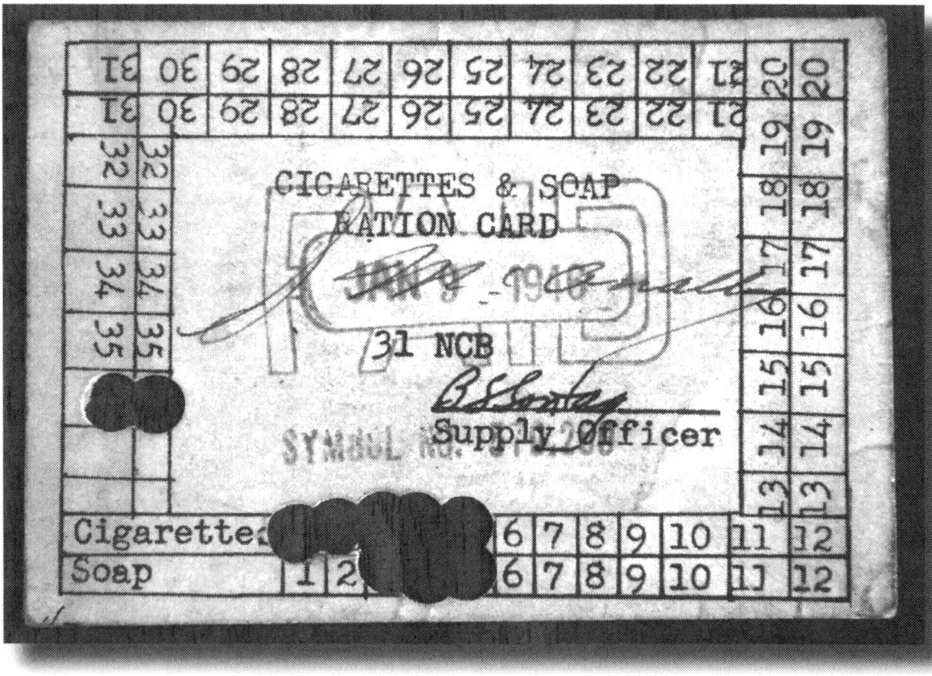
CIGARETTES & SOAP
RATION CARD
JAN 9 1946
31 NCB
Supply Officer
Cigarettes
Soap

Image 30:
One of his ration cards from early 1946.

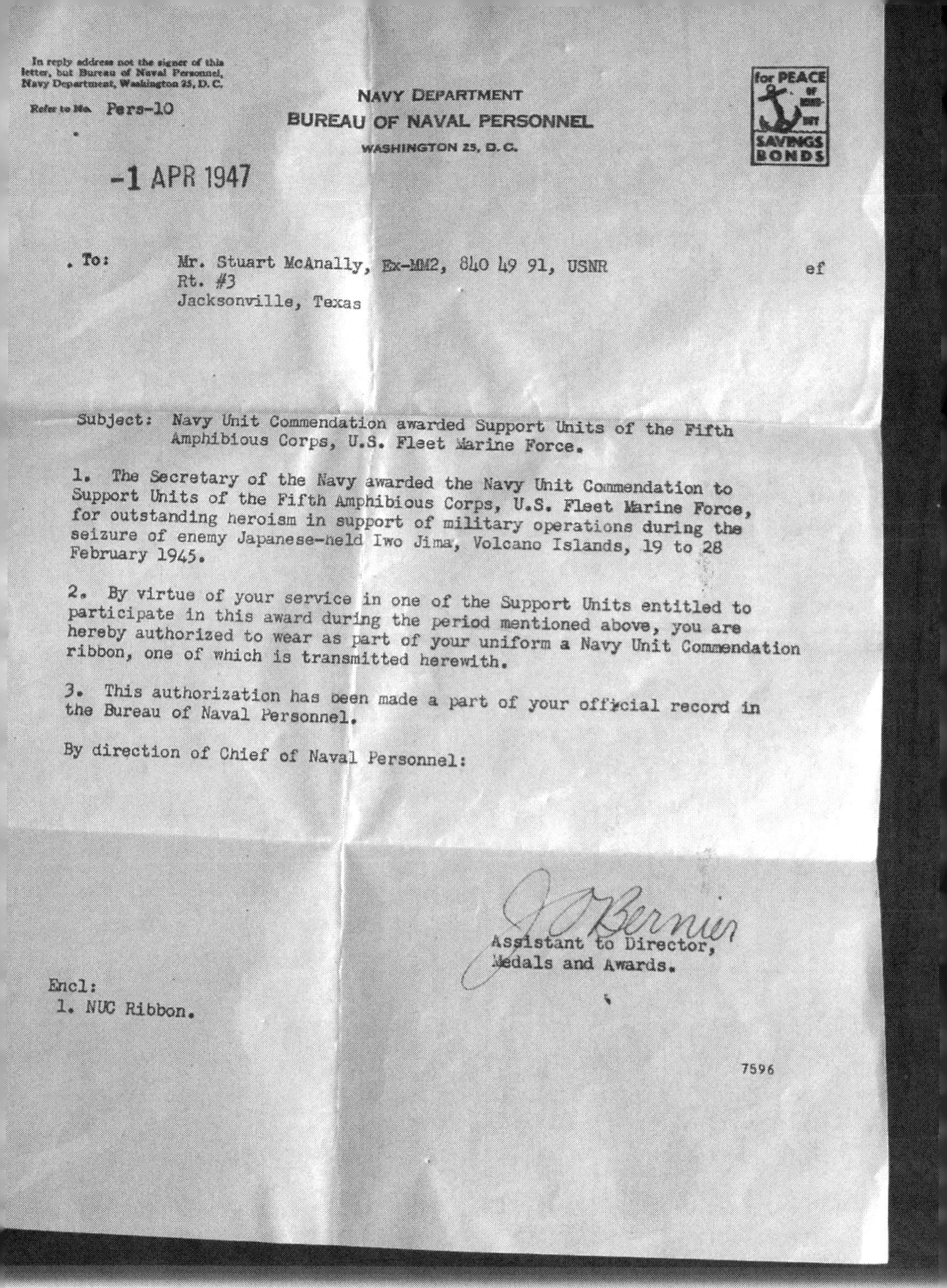

In reply address not the signer of this letter, but Bureau of Naval Personnel, Navy Department, Washington 25, D. C.

Refer to No. Pers-10

NAVY DEPARTMENT
BUREAU OF NAVAL PERSONNEL
WASHINGTON 25, D. C.

for PEACE and FREEDOM BUY SAVINGS BONDS

-1 APR 1947

. To: Mr. Stuart McAnally, Ex-MM2, 840 49 91, USNR ef
Rt. #3
Jacksonville, Texas

Subject: Navy Unit Commendation awarded Support Units of the Fifth Amphibious Corps, U.S. Fleet Marine Force.

1. The Secretary of the Navy awarded the Navy Unit Commendation to Support Units of the Fifth Amphibious Corps, U.S. Fleet Marine Force, for outstanding heroism in support of military operations during the seizure of enemy Japanese-held Iwo Jima, Volcano Islands, 19 to 28 February 1945.

2. By virtue of your service in one of the Support Units entitled to participate in this award during the period mentioned above, you are hereby authorized to wear as part of your uniform a Navy Unit Commendation ribbon, one of which is transmitted herewith.

3. This authorization has been made a part of your official record in the Bureau of Naval Personnel.

By direction of Chief of Naval Personnel:

J. O. Bernier
Assistant to Director,
Medals and Awards.

Encl:
1. NUC Ribbon.

7596

Image 29:
The letter awarding the Navy Unit Commendation for Dad's service with the 5th Marine Division during the battle for Iwo Jima.

Photo xxix

Image 31:
A clipping Dad kept and made a note on from Iwo Jima.

Image 32:
A Japanese newspaper Dad picked up outside one of the many caves on Iwo Jima.

Photo xxx

Image 35:
Postcard of Hiroshima, Japan after the atomic bomb was dropped.

Image 34:
Photo of devastation at Nagasaki, Japan.

工場へ激勵電
東條軍需相から

實戰防空だ"那覇の教訓"で

爆弾より火災の惨
初期必消へ鍛へよ隣組

陸鷲必殺の翼
る隊長機以下を

他力本願

Image 33:
Part of Japanese newspaper Dad sent back home with notation

Photo xxxii

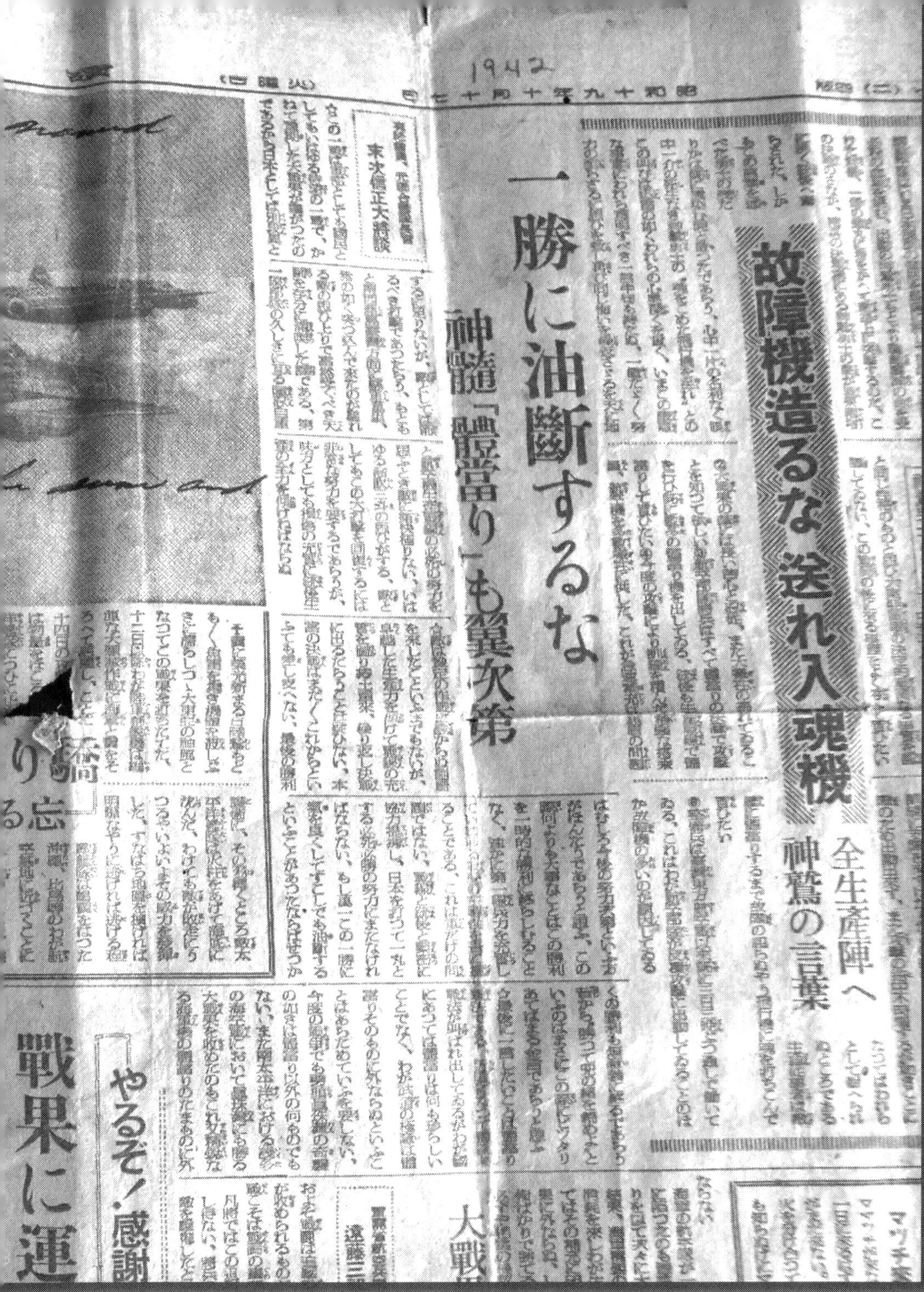
1942

昭和十九年十月十七日（火曜日）

故障機造るな 送れ入魂機

全生産陣へ 神鷲の言葉

一勝に油斷するな

神髄「體當り」も翼次第

戰果に運

やるぞ！感謝

大戰

Photo xxxiii

⇐Image 36:
Dad and a couple of fellow servicemen at Omura, Occupied Japan, taken by a Japanese photographer – circa late 1945 or early 1946.

Image 37:
Mom and Dad soon after they married in April 1947.

Image 38:
Dad trying on his Navy uniform in 1974, thirty years after he joined. It still fit!

Image 39:
The last picture taken of all five of us siblings on Mother's Day 1985. Left to right: Me, Josh, Melanie, Sherry, and Shaun.

Image 41:
Here we are at an anniversary celebration for Mom and Dad in 2013. Front: Melanie, Dad, Mom and Sherry. Back row: me and Shaun. Being together is a priority with our family.

Image 40:
Veterans of the 31st CBs at reunion in 2002 - Back row left to right: George Young, Ben Sontag, Clifford Paxton, Martin Devries, Stuart McAnally, and Eldon Robbins.

Photo xxxviii

Front row left to right: Jim Dycus, Paul Marabito, Bill Patrick, Jim Chisholm, Bernard Duffe, and John Kelly.

Photo xxxix

Image 41:

Dad looking at his 31st CB book and the small journal he kept during the war – 2018.

Photo xl

Why did they ask me? I'm only twelve years old.

I've seen death up close. When my little brother Ebby died, I remember they put nickels on his eyes to keep them closed. I thought when you died, you closed them yourself. To find out they would just keep staring out, looking at nothing, was a little scary for me. After all, I wasn't even ten yet when he died.

Now, they want me to go help dig a grave for Mrs. Buffalo. I didn't really know her that well even though I'd been to their house with Sis, my oldest sister, and her husband Leo. Mrs. Buffalo was his Mama.

She was always kind to me. I'll never forget spending the night at her house once when I was about six years old. It was pretty cold, and she got some blankets for me to sleep with. I'd been used to bulky quilts, but those blankets were like something from heaven. They were so soft and warm. I'd never felt anything like them. I knew they weren't homemade...the Buffaloes must be rich, I had thought.

Now she's dead, and they want me to help dig her grave. Saying no is not an option when you are twelve and you've been raised to mind, so I go with them. We end up at the top of the red dirt hill, next to where Grandpa John Ryan and Grandma Tabby Jane are buried.

Since Mrs. Buffalo wasn't really blood kin, I wonder why they want to bury her so close to our folks. Then, I see the graves of more Buffaloes - and a Miller that someone says was her sister - so I feel more settled about it.

Digging is hard in this packed red clay...guess that's why they want a strong back like mine. Three of us do most of the digging. Once we get down about two feet and it gets harder to maneuver in that narrow space, we take turns. Finally, we get it deep enough to satisfy Mr. Buffalo.

They say something, but I can't hear it over the roaring in my ears. I see one of the men motion to lower the box into the ground. I do hear the box though, when it lands with a thump at the bottom of the hole. As we start shoveling the dirt back in on top of her wooden coffin, I can hardly bring myself to do it. I don't know why that bothers me so much…hearing that sound, hearing those clods hit her coffin, but it will haunt me for the rest of my life.

I feel someone shaking me, but I don't know why.

"INCOMING!" someone yells.

It sounds far away…sssthump!…sssthump! I feel the clods of dirt hitting me, and I want to yell for someone to get me out of this coffin. I can't breathe.

"Tex! Tex! Get down!" I hear Peterson yell as he yanks on my arm.

I wake up and realize mortar fire is exploding all around us and I'm covered with volcanic ash.

This foxhole feels like a coffin…all covered in twisted metal, hot and claustrophobic. I just want to run…run home, away from the falling clods, away from death, away from this hell hole, but I know there's nowhere to run.

Chapter Six

The days were getting shorter and the leaves had almost all left their stations, thanks to an early November freeze. Thanksgiving was just two days away, and already my sisters were busy in Mom and Dad's warm knotty pine kitchen preparing for our annual family feast. Usually we had anywhere from 25 to 30 relatives join us for the holidays, making it a loud and happy time.

Melanie, the oldest, has always made the chicken and dressing, at least since she got married. She also makes her famous yeast rolls for family get-togethers, making as many as five or six dozen for special occasions. Those two items are always the highlight of our holiday meals.

She and my sister Sherry were making pies - pecan, pineapple cream, chocolate meringue, and raisin - mine and Dad's favorite. Dad and I sat at the kitchen table watching the frantic activity when he turned to me.

"It's getting pretty thick in here," he said.

I grinned and nodded towards Melanie digging through the lower cabinets for a particular pan, clanging and rattling them something awful.

"And loud too," I added.

"Guess we'd better clear out." He looked over at Melanie dragging pans out on the floor as Sherry cranked up the mixer to a deafening roar.

"Want to go to the sunroom?" I asked.

"We can take a look at my stuff in that old wooden box if you want. I remember you asked about it before," he said as he rose from the table with his cup and plate.

"Sure, I'd love to." I followed him to the sink with my cup, dodging Sherry on her way to the stove to check on a piecrust. After adding our dishes to the ever-growing pile, we made our escape to the relative quiet of the sunroom. It's a favorite place to sit in the fall and winter, a cozy retreat to admire the view when the weather is too cool to sit outside. The spacious room has multiple floor-to-ceiling windows that provide a 180° panorama of the place we all love. When we got to the sunroom, we turned to the left towards the only wall without a window.

"Did I ever show you the letter I got from President Truman?" Dad asked as he opened the closet doors and rummaged around on the top shelf.

"No, I don't think so."

"It's in here, I believe." He reached up and grabbed the metal handle of a battered wooden box. As he pulled it down towards him, he reached up to support the bottom of it, keeping it level so the things inside wouldn't get too scrambled. I moved some magazines off the table next to his chair and he set the box on it, then flipped up the two worn metal latches on either side, and slowly opened the lid.

"Wooo!" he said. "It's a mess of papers in here. I hope I can find it."

I peered inside. "How long has it been since you looked through all this stuff?"

"Oh...I don't know, several years."

I spied some postcards and letters in a plastic baggie, picked it up and turned it around.

"I thought you said you threw away all of the letters you got, except the one from Grandma." I held out the plastic baggie for him to see. "Then what are these?"

Dad took the bag, picked up his glasses off the table and put them on to take a closer look. He squinted at the bag and turned it over.

"These look like the letters Mama kept...the ones I sent her and Babe." (Babe was what everyone called Dad's older sister, Edith. During the war, Babe worked at a ship-building plant down on the Texas coast. She was one of the original 'Rosie the Riveters.')

"How long has it been since you read them?" I asked. Dad looked up, still holding the bag aloft. "I don't know that I've ever read them, not since I put them in this box."

I stared at him. "And just how long ago was that?"

"Oh...thirty-something years ago I guess. Your Aunt Babe gave them to me after Mama died."

"Do you mind if I read them?"

"No," he said as he handed back the plastic bag. "But I'm afraid they'll be pretty boring. We weren't allowed to write much detail. All of our mail during the war was censored."

I smiled as I took the bag and opened the top. "It will still be neat to read what you wrote back then."

"Do whatever you like," he said. "There are no great secrets in there... that I know of." He looked at me over the top of his glasses, which had slid down his nose a bit. "You know, I haven't looked through that bag in so long, you might find something I've forgotten about or overlooked. You're welcome to look through it."

Dad turned to sit down in his chair and let out a low 'whew.' "I think I'll sit down while I look for that letter from Truman."

He reached over into the box on the table beside him and picked up a bundle of papers. He placed them on his right leg and sifted through them one by one, placing the ones he had looked at on his left knee.

I sat down in a wicker rocker next to the bank of windows, then carefully pulled out a few of the yellowed envelopes and postcards and began reading. Several minutes passed in silence as we each got lost in the past.

"Here it is," he said. "I knew it was in here somewhere."

He held out an ivory-colored paper for me to see. It had the official White House seal at the top and was in surprisingly good shape for not being in a protective envelope. It was a letter of thanks for his service during World War II. I'm sure all of the servicemen received one, but it was still pretty cool to see something that was a part of history.

When I finished reading it, I gave it back to him and he slipped it into the pile he had finished looking through. He picked up the stack on his left leg, set it aside, and reached in the box for more papers.

"This is fun," I said as I went back to the rocker and sat down. "It's like a treasure hunt."

Dad smiled as he looked at a faded postcard. "I reckon so" was all he said.

I didn't think he considered all this stuff too important...but it's priceless to me. Every scrap of paper and keepsake in that old wooden box is like a window into the past. It's not only our family history, I thought, but a historical representation of that time in our nation's history. I felt like I should have on white gloves to handle those delicate documents. I tried to be careful as I opened the brittle envelopes slowly and removed the faded letters inside. Though mostly written in pencil, I could decipher some of them pretty well, even though Dad's scrawling lettering was a bit hard to read.

As I reached into the bag again, I felt something small stuck in one corner. I stretched the bag open where I could get to it and pulled out the tiniest little notebook, yellowed and without a cover.

"What's this?" I got up and handed it to him.

He stared at the small lined pages. "This is the book I kept during the war. It did have a blue cover; I guess it got torn off. It was for addresses. I wrote some of the guys' names and addresses in here. Some of them wrote their own addresses in it…and some of the girls I met too," he said as he thumbed through it.

"Can I see it?" I said.

"Not much to it," Dad said as he handed it back to me. "I did manage to keep it all through the war – I guess that's something."

As I read the entries one by one, I saw familiar names: Allen Locket, Jerry Seiberlich, and Uncle Monroe's address when he was stationed at Fort Sill, Oklahoma. I smiled when I saw Lou Renzo's name. He never called on her again, but he did get her address. Then I turned past several blank pages and came upon a page of writing with dates on it. I turned another page, then another, and realized I was looking at a short journal of Dad's service time.

"Dad!" I practically shouted. "Do you remember writing a journal of when and where you went while you were in the service?" He looked up at me in surprise, then his brows knitted together as he searched his memory.

"I remember writing some stuff down. We weren't supposed to, but I did anyway. Why?" He gestured towards the little book in my hand. "Is that where I kept it?"

I waved it at him in jubilation. "It sure is! This is great! I'm so glad you did write it down. This tells exactly where you were, when you left and everything."

I turned the pages and scanned all the entries. "This follows you from the time you left home, through boot camp, Endicott, California, Hawaii, everything…right up to when you went ashore at Iwo on February 21st, 1945!"

I couldn't believe what I held in my hands – actual documentation of the timeline of events he had described all these years. I was ecstatic. Dad was amused.

I ran to the kitchen to tell my sisters about the discovery. They stopped their cooking and dishwashing to take a look. We were all in awe that he had managed to keep a record of his movements and keep up with such a tiny notebook, even under such severe circumstances.

A timer went off for the pecan pie and forced Melanie and Sherry to turn their attention back to the pie-making. I went back to help Dad look through the rest of the box. But I was sure of one thing, that little book was the *find of the day*.

We spent a good hour or so going through the rest of the contents of the box…I listened while he reminisced about the souvenirs he'd brought back with him – a black and white photograph of a very pretty Japanese girl, boxes of Seabee matches, a couple of Japanese match boxes he had picked up on Iwo Jima, some Japanese money, a silk handkerchief decorated with cherry blossoms and a Japanese house, a pair of beautiful wooden chopsticks.

As we worked our way to the bottom of the box, Dad became more animated and I could see that he was enjoying this trip down memory lane – for the most part. He has a unique way of making the best of any situation. I am sure that ability served him well on Iwo as it has since then. While we looked through his relics from that long ago time, I could see a sadness wash over his face now and then, but it was so brief, that if I had not been looking at him, I would have missed it.

I remembered asking Dad once how he dealt with seeing all that death and misery on Iwo. He got real quiet before he answered.

"Some things, you try not to think about too much – you just have to put them behind you and move on or it would drive you mad. War is a terrible thing. There is nothing that can change a man the way war can – for better or worse. You just have to make your mind up which one it will be."

As I watched him today, I knew I didn't have to ask which one he had chosen. He had chosen life, a better life…and I was glad.

Later that day, after all the cooking and massive clean-up, we gathered around the fireplace to visit. Melanie and Sherry didn't get to hear some of Dad's earlier reminiscing, so they asked him to fill them in. We talked about the little notebook and all the things we found in his keepsake box. I knew we would all be busy the next few days with more cooking, company, and such. It might be a long time before I would get to ask him about Iwo again, so I took advantage of that quiet evening hour and asked him to tell us more.

"Do you mind going back in time again…to Iwo? I asked as we settled into our respective chairs. I won't forget his unexpected answer.

"It's never far from my thoughts."

Even after almost seventy-five years, that time held such a grip on his mind it was startling to me.

He looked over at Melanie and Sherry. "Y'all don't want to hear about this stuff, do you?" Both of them nodded. Dad looked over at me. "You did leave me in that foxhole, didn't you?"

I grinned at him. "Guilty as charged."

"Well," he said "I'd better get myself out of there then. Let's see..."

He leaned back in his recliner with both hands on the arm rests. Then he stared into the fire as he continued the rest of the story.

"Like I said, it was chaos those first few days. I remember the first night was pretty scary. I don't remember sleeping. You could hear those big shells coming in. They made a whistlin' noise. One night, one screamed pretty close and something hit at our feet where we crawled in. We had a makeshift roof on our hole out of the mangled steel mats that were scattered on the beach." He motioned with his hands over his head. "Anyway, when it got daylight, we saw a big hunk of that shell right where we crawled in. Pete and I were very fortunate. Several times, shells hit foxholes nearby, killin' the Marines inside. We had some close calls, but were never hit. The shelling of the beach went on for days and days."

"That was a close call," I said. "What did y'all do for food and water while you were hunkered down?"

"When we landed, we had full packs with a blanket, a gas mask, two or three boxes of rations, and a canteen of water. We also had our semi-automatic rifles and a magazine that held fifteen rounds. We also wore our duty belt that had our big G.I. knife on it. It was about seventy pounds of gear in all and if it got wet, it was pretty heavy.

"Anyway...after several days in the foxhole, Peterson and I ran out of food. We went down to where the Marines had unloaded a stack of rations from the ship, but the Marine in charge wouldn't let us have any food without an order. A lot of people don't know this, but on Iwo we were all dressed alike; us Seabees looked just like Marines. I couldn't imagine them refusing rations to anyone, especially given the circumstances, but I guess he had his orders.

"Anyhow, I went down the beach a ways, out of their sight, and found a piece of paper, brown wrapping paper, nothing official, and a pen. I wrote out an order for us some rations and signed it 'Officer in Charge.' I was a Petty Officer, and I was in charge of me and Pete's food." He looked over and grinned at us.

"When I went back to the Marine and gave him that slip of paper," the corners of Dad's eyes crinkled, "we got a case of rations. We had it made, at least as far as food went. The Seabees didn't have any food of their own because we weren't even supposed to be on the island yet. They had to bring us in to get their equipment off the beach. We didn't have any supplies or anything of our own until later when the rest of our group came ashore. I remember while we were still living in our foxholes that the galley cook got ahold of some flour and made us some biscuits. They had a stove right out in the open on the beach. We got two biscuits apiece. It was the first real food we'd had in weeks.

"Later on, I was helping unload supplies and I *procured* a case of sliced pineapple, a can of syrup, and a box of bacon. I ate the pineapple… almost made myself sick on it. They were in big cans, No. 3 cans, I think. I'd eat a can at a time…I fared pretty good. After a while, I finally gave our cook the syrup and the bacon. I guess he wondered what I was doing with it, but he didn't ask any questions." He grinned.

"Early on, workin' on the beach, we were under Marine command. I remember this Marine officer prancing up and down the beach. I can still see him to this day. He had a pistol on his side and a belt of bullets around his waist and going up over his shoulder. He cut quite a figure. He pranced up and down that beach a hollerin' orders. He told us not to 'hit the deck' until he gave the order.

"What gave us a break, the Japanese would run their guns out on the rails they had built and shoot, then pull them back into the holes again, giving us a little time to do something. One time I was trying to help hold this landing craft. The water was rough and when I hit the deck the wave would come up over me, but I was keeping my head down. Then the officer would give the order and we'd jump up to work.

"But we had one fella that when the shells got to bursting, he couldn't stay on his feet. Every time the shells started popping he went to the ground. The Marine officer finally told him, 'You hit the deck early one more time and I'm sending you to the front and you're gonna stay until you don't come back.' Of course, it was kind of an empty threat because on Iwo the Japanese could hit you from anywhere on the island. It didn't matter if you were on the front line, the beach, the interior - on Iwo, there were no safe zones."

I thought about having to 'hit the deck,' as he called it. I didn't see how getting down would keep you from getting hit by the shrapnel since it went in all directions, so I asked him about it.

"Well," he said as he motioned with his hands, "when a mortar shell explodes it goes up and out, so if you are lying flat on the ground, most of the time it will go up and over you - unless of course, you take a direct hit."

I nodded. Now I understood a bit better how important it was for their safety to get down flat on the ground. And I could sympathize with that guy that kept hitting the deck. I might not have been able to stay on my feet either.

"One of the most dangerous things we encountered as we tried to move about the island was booby traps and land mines." Dad continued. " The Japanese had lots of booby traps set on Iwo and we were warned before we landed that if we found a pocket knife, a watch, or anything, not to touch it. The Japanese had several different types of mines and one of our demo guys got hurt, but not one of them was killed on the demolition squad…another miracle. Holliday and 'Coon' Cecil Watkins were part of that group. I learned later that those guys removed thirty truckloads of bombs, duds, and ammunition from the island. They did a great job of clearing the island so we could do our work."

"Tell us about when you saw the flag go up," Melanie said.

"I was on the beach when I saw the flag go up on the fifth day. We were working on clearing the beach when I looked up and saw some of our guys scurrying up the side of Suribaci. I knew it was our guys because they were out in the open. In just a few minutes, I saw the flag go up; it must have been the first flag, and then I went back to work. What it meant to me was they weren't going to be shooting at us from the top of Suribaci anymore…we hoped.

"There were still Japanese up there in holes, but that quieted it down some…at least on that end. Martin said that the ships all blew their horns and carried on when they saw the flag go up from out there, but I don't clearly remember that…I was too busy workin' and trying to keep from getting shot. A lot of people don't realize that there were two different flag-raisings on Iwo. The first one was with a smaller flag that went up about mid-mornin', but I guess the brass didn't think it was big enough, because they ordered another, larger flag to be raised and that was when the second flag-raising took place. It probably took a while to get one from the ship and back to the island. I don't remember seeing the second flag go up; I had my mind on my work. I don't think any of us had an idea that event would come to symbolize the war in the Pacific and become such a famous symbol of World War II…it was just something that happened while we were trying to survive and do our job. I later heard at one of our Iwo Jima Survivors meetings that a Marine photographer Louis Lowery had taken the photograph of the first flag-raising, but not many people remember him…they only know about Rosenthal and his famous picture."

"That's true," I agreed, nodding my head. "Most people I've talked to about Iwo don't know about the first flag-raising."

Sherry got up to get her and Mel some pecan pie; they claimed they wanted to taste test it before everyone arrived tomorrow for Thanksgiving. We waited till she got back, then Dad picked up his story.

"It was in the first week that I saw the first of our officers, Ensign Pierce. He was a class act, a real gentleman. I liked being under his command. We were under Marine command until the rest of our bunch, the 31st, finally all came ashore. When the rest of our outfit came in, we moved from those first foxholes on Red Beach where we landed, over to the west side of the island, they called it West Beach, and dug in there. It wasn't very wide at that end of the island. We had to stay in those holes until they took the whole island because we were going to build our camp on the north end. We were in those foxholes about two months. It was mid-April before we moved to a permanent camp area and got to set up tents. Me and Clarence Peterson were in a foxhole together for those two months. I don't remember pullin' off my boots during that time. I might have, but I don't remember. We were a dirty, gritty mess, but we were still alive.

"Even though we moved into tents, the caves were still occupied by the Japanese, and we had to maintain a perimeter guard for our safety, but after spending those hellish two months in a foxhole, those tents were like upgrading to a fancy hotel."

He paused for a minute, lost in his thoughts.

"You know…Pete and I got separated when we went to Japan and when I got back to California, I tried to find him. I called every Peterson in the Greater LA phone book, but I never found him. I thought he might have married Mary and stayed in California. I never did get in touch with him…I always regretted not being able to see him again."

Sherry said, "I'll bet you could have found him if you'd had the Internet back then."

I laughed and said, "Yeah…that's about all technology's good for, unless you don't want to be found."

Dad smiled. "Yep, things were a lot different back then."

"Anyway," I said, anxious to hear more, "back to your story."

He straightened up in his recliner and stared into the fire for a bit before he began.

"After we got the equipment off the beach, they ordered us to start to clear the airstrip. The way that No. 1 airstrip was built, it was short, just for fighter planes. The runway went north and south from Suribaci. Back on the west side, there was a tarmac, or a place for the planes to park. In the tarmac area, that L-shape, there was a knob up pretty high. Ensign Pierce and about twelve or fifteen of us headed up there, but the Japanese shot us off. That's how close they were yet…they opened up on us and we had to run back over the hump and take cover. They were shooting at us from the knob that was beside the airstrip, near where they would park the planes. We didn't even get started workin' before they started shooting. Pierce said, 'Well…we'll have to wait awhile'."

"Steed, a guy from Alabama, got shot in the leg before we could get off the airstrip – that ended his time on Iwo. He was hit in the leg while we were running for cover. He was about 18 or 19 years old. Years later, I got a letter from his daughter, tellin' me he had died. She said he had trouble with that leg the rest of his life."

Dad shifted in his recliner and lowered the footrest. "Looks like that fire needs a little attention," he said as he got up and reached for the iron poker. We all watched as he sent a shower of bright embers up the chimney and then added another log to the small stack of burning red oak.

"You know…" he said as he poked the log into place, "the Japanese were hunkered down in those caves for the duration. They had no intentions of ever making it off that island alive. It was considered a disgrace to be captured – especially when all your fellow soldiers had fallen in battle. The Japanese had a code of honor and respect that ran deeper than their desire to live…no Jap soldier worth his salt survived a battle like Iwo; he went down fighting…that's what we were up against." He stepped back from the hearth, put the fire poker in its stand, and set the screen back into place.

"What else happened on Iwo, Dad?" Sherry asked.

He turned to face us and stood up straight, his hands behind his back, warming himself by the renewed flames.

"Well…" Dad paused, "lots of things probably not worth tellin' and some I would rather forget." He walked over to his recliner and settled back before he began.

He didn't usually talk much about the harder things he saw and dealt with on Iwo, but tonight, in the glow of the firelight, he told us one of the things that he couldn't shake…even after more than seventy plus years.

"We would gather up the bodies and row them up in between barrages of gun fire from the Japanese. We couldn't move them very far, we didn't have much time, but we put them in rows, out of the way of our machines, and covered them with blankets. We had a row just below our foxhole. I tried not to think about those boys under the blankets... but it was hard. It took a long time to round up all the dead. The first few days, we gathered up the wounded, and as a boat came in, we would get them loaded so they could return to the hospital ships. We only had that short window between barrages from the Japanese to move the wounded and row up the dead. It was awhile before we could do anything with the bodies, there was no time to bury them. It was pretty cool that time of year, so that helped, but the smell was hard to take. We were surrounded by the dead, and the sounds of the dying were all around us...it was a terrible thing. I remember hearing one young Marine calling for his Mama...then he went quiet."

Dad didn't speak again for several minutes. He looked so sad as he unconsciously rubbed his hands round and round on the arms of his recliner, as if trying to smooth something away.

He finally continued. "All of those young men, cut down in their prime. I think that was the hardest part...thinking about all those young men wasting away on that island, so far from home. It was hard to shake loose from...lookin' in the faces of those lifeless young men...knowing they were some mama and daddy's son, maybe some kid's dad. There were so many...and it was like that all up and down that beach."

He shook his head. "There were some Marines in a foxhole just above us that didn't get up one mornin'. They checked and the Japanese had eased in and cut their throats during the night. Iwo was a dangerous place to be bedded down.

"The next mornin', they had the bodies rowed up and covered with blankets. Chaplain Gregory was going through the row, checking their dog tags and found a Japanese soldier hiding among the dead Marines. The Jap was planning on making the day there, lying among the bodies and then slip out when night came and kill some more. I heard that the Chaplain just shot him. All of the Marine officers carried pistols. We all had to be on alert for Japanese behind our lines...that was their intent, to kill as many of us as they could before they died.

"The Japanese slipping out of their holes at night was a constant thing. They had tunnels all in the island, places for supplies and for them to 'hole-up.' They were prepared for a fight...a long one, like I said. I believe if we hadn't dropped the super bomb, we would have been in trouble because we would have had to take Japan one foot at a time. It would have been tough and I believe a lot more people would have died...on both sides."

Dad ran his hand through his white hair and sighed before continuing.

"We had GI dogs that could smell them out and the Japs were afraid of them, but we lost a few dogs. One of our guys was on guard duty one night near the beach and shot something coming down the beach. Turns out, it was one of our dogs from the canine unit. I guess it had gotten loose and was exploring the beach. That poor fellow felt really bad, but we had our orders to shoot at anything that moved."

His face lit up at a memory. "There was this fellow from South Texas named Wassom - he was a crack shot. One night, he was on guard duty and saw something moving down a slope ahead of him. No one had any kind of light except certain officers, and like I said, we were under orders to shoot anything that moved after dark; that's why we were under orders to stay in our foxholes. Wassom opened fire on the moving object, and the officer in charge came to investigate. He found that Wassom had unloaded all fifteen rounds into a newspaper!" We all laughed.

"Sometimes, to keep the Japanese from infiltrating our lines, we would send up flares. They had parachutes on them that would open up and a bright light would shine as they floated down. Those flares would light up the whole area."

He shifted in his chair and pulled up the footrest before continuing.

"One night I was on guard duty and I got the fright of my life. It was one of those pitch black nights, no moon, no stars…one of those thick nights when you can't see your hand in front of your face, like being in a cave. I had no light; there were no flares, so I had to rely on my hearing and other senses. It was about midway through my watch when I heard it…something was moving towards me. Softly, hardly making a sound, it came closer and closer. I strained with every muscle and nerve to try to determine the source. If it was a man, he was being very quiet. The Japanese were experts at sneak attacks; I knew they could be out there, but I wasn't sure, so I just froze. It came closer and closer…I thought if I couldn't see them, maybe they couldn't see me if I just stood still. Finally, it was right next to me…I still didn't move. Then, it touched my leg…and it purred. It was a big black cat! I was so relieved; I would have laughed if I hadn't been on guard duty, but the possibility that there were others out there, not so furry or friendly, stayed on my mind the rest of the night."

We all got a chuckle out of that story. My mind was still full of questions - I just had to ask one I had often wondered about.

"What did y'all do with the Japanese dead?" I finally asked.

"The Japanese pulled their dead back into the caves as long as they could. I think it was a psychological ploy so it would look like we weren't killing any of them, at least it would look that way. Towards the last, they didn't have the opportunity to do that though. The Japanese dead found by our troops were hauled off and put into a big pit, and covered for health reasons.

"We had to take care of things the best we could under the circumstances, but nothing about it was pleasant. I remember one time I was eating in our foxhole and to the left of me, I could see a pair of legs sticking up. I knew it was a Japanese soldier because of the way the leggings were wrapped. They wrapped them round and round up to the knee," Dad motioned with his hands as he talked, "our leggings laced up. Kinda took away your appetite, eating next to that sort of thing.

"I don't know why that stuck in my mind." Dad shook his head. "It's amazing what you can get used to when you are in a situation like that."

My stomach lurched at the thought of trying to eat surrounded by so much death. I wondered at the ability of those men to keep their sanity…and their appetites. From the horrified expressions on my sisters' faces, I'm sure they felt the same way.

"Our job was to get the No. 1 airstrip operational as soon as possible." He continued. "There were holes as big as a truck…bomb craters that we had to fill in. We managed to get the airstrip where planes could land in just a few days. We kept workin' on it for a long while, improving the airstrip so it was safer. Of course, there was the constant upkeep and maintenance we had to do too. The Seabees didn't have all of the equipment we needed there on the island, so we had to build it or fashion a makeshift machine to do the job. What some of those mechanical fellas could do was really amazing. The Seabee motto "We Build, We Fight" and our "Can Do" attitude went a long way in making our time on Iwo a success.

"They said the purpose of taking the island was to give crippled B-29s that couldn't make it back to Saipan, a place to land, and for the fighter planes that escort the B-29s to have a place to come and go from. I don't know for sure if the Enola Gay ever landed on Iwo Jima, but I do know that we had dug a pit in case it needed to land and change the bomb to another plane if need be."

Dad tried to stifle a yawn. I suddenly felt tired and decided I needed to go on home and get some Thanksgiving preparations finished before bed. I said my goodnights with hugs all around and left the three of them sitting in front of the fire.

On the way home, I couldn't help but think about all those young men who lost their lives on Iwo Jima, and how thankful I am that my Dad wasn't one of them.

My eyes feel all prickly and hot, like pine needles are sticking in them. I can't breathe. I hear Dad shouting and I feel someone pulling on my arm. I see flames licking up the doorframe like some big ole orange tom cat.

Gray shadows pull me into the cold night air. Everything sounds far away. I hear the scraping of something heavy being drug across the rough dogtrot floorboards and I hear Mama calling, "Are they all out?"

Somehow, I find myself about a hundred feet from the burning house, holding Floyd's hand. We are all standing here, me and Floyd, Mama, Dad, Sis, Babe, Eva and Josie, staring at our house as it collapses with hardly a sound into a pile of red hot coals and mangled tin.

I look over and see Mama holding baby Monroe, and she's crying. Dad has his arm around her shoulder.

We all got out, with just our night clothes. Then I spy Grandma's trunk in the grass about ten feet in front of us. That must have been what I heard scraping across the dogtrot...Dad always manages to save the trunk.

I am glad he does.

It is all Mama has left now...except us.

CHAPTER SEVEN

Last night I dreamed I couldn't find Dad. I looked everywhere. I finally woke up, but like in some dreams, I knew I was still asleep...but it seemed so real.

I awoke to a darkness that was so thick it felt like a velveteen blanket, the kind that if it settles around your nose, you feel like you are smothering.

I felt a heat emanating from somewhere underneath me and the smell of sulphur filled the air. Crouching in what I guessed was a small cave or a hole, I froze, unable to move for what seemed like hours.

Suddenly the rat-tat-tat of gunfire sounded in the distance and it was answered on either side of me by the crack of rifles. In a brief flash of fire, I saw men's dirty streaked faces anxiously staring out into the night.

My heart leaped as I instantly recognized the one closest to me...it was Daddy! He was only about six or seven feet away. Not my white-haired ninety-five year old Dad; this one was young. The high cheekbones, strong jaw and profile were the same; it had to be him.

My heart was pounding...I had found him. But when I tried to move closer and call out, I was helplessly frozen in place, unable to utter a sound.

More gunfire erupted, and Daddy and the other two men returned fire. I could smell the acrid smoke and see the dust clouds in the flashes of light.

I wanted to crawl to where he was...protect him...keep him safe. I wanted to take him away from this horrible place, but all I could do was lie there, listening to the intermittent exchange of rifle fire, watching for my Daddy's face in the flash.

When I finally woke up...for real, I was crying. It took me a minute to orient myself to the present. The clock on my nightstand convinced me I was indeed in my own bed.

It was the first time I'd dreamed of Dad and the war. I guess it was prompted by our talk last night in front of the fire.

As the pounding of my heart slowed to a more normal rhythm, I thought of all the times Dad had protected me growing up… physically, emotionally, and spiritually. Now, I want to protect him, save him from the enemy I know is coming. But I know it's not possible.

• •

Later that morning, still feeling vaguely unsettled and needing a lift, I decided to go over to Mom and Dad's and see what he was up to. I found Dad up at his big metal equipment and storage barn on the hill northwest of the house, sorting through some old lumber and choosing pieces to use to build two 'new' feed troughs. The two in the pasture had 'seen their better days,' as he says, and he decided that before they fell completely apart, he'd better replace them. Dad had made noise about this project for a couple of months, but with winter about to set in, he'd decided to get started.

"Well, did you find any more mice in your feed room?" I asked as I walked into the barn.

"Good Morning!" Dad said as he straightened up and pulled off his gloves. "Saw one yesterday. He had climbed over in a five gallon bucket. I found a two by four and ended his earthly pilgrimage."

I smiled. He had a way of saying things that made everything he did sound out of the ordinary, even when he was exterminating rodents. He was never disrespectful to any life form, no matter how much they tormented him. Gophers, moles, rats, mice…those he usually buried or let us give to our cats, circle of life sort of thing, I suppose. Some people kill just for the pleasure of killing, but not Dad, only when he deems it necessary, and even then with a certain reluctance. I imagine it stems in part from seeing so much death and senseless destruction during the war. War does that I think, makes people look at things differently, sometimes for the better…sometimes not.

"Well, I hope no more get into the feed room, especially since you just got that load of range cubes and mineral," I said as I pulled out a five-gallon bucket, flipped it over and sat down.

"Yeah, they sure can make a mess."

Dad took a seat in the folding chair he keeps at the barn for when he 'takes a notion to sit a spell.'

Taking off his cap and placing on his right knee, he leaned back in the chair and let out a loud 'Whew.' Mom said he had been up there awhile, and I guess he was ready for a break when I showed up.

Whenever I get a chance, I ask Dad questions about his childhood, service time, or his painting career. Sometimes the stories are ones he's told dozens of times, but every once in a while, he'll surprise me with something I haven't heard before or maybe didn't remember. Today was one of those days.

"I know you get tired of me asking about Iwo sometimes, but I want to know all you can remember. I think it's important…it's part of our history." I said as we sat in the shaft of sunlight thrown across the barn from the open door.

He gave me one of his amused looks and said, "Well…at least somebody's interested in the old days."

"All of us kids are interested in them, Dad, you know that," I could tell he was teasing, but I wanted to encourage him to keep talking.

"What do you want to hear about today?" he asked.

"Tell about some of the things that happened while you were working on the airstrip or anything you can remember about other service men you saw while you were on the island." I leaned forward, elbows on my knees, and waited.

Dad was facing the barns' partially open sliding door. I followed his gaze and noticed the countless dust motes that sparkled in the shaft of sunlight coming through the opening. It was several minutes before he started talking.

"Well…getting the Number One airstrip usable was our first priority. Fixing all those huge bomb craters was a tall order. Problem was, the volcanic ash on that end of the island wasn't any good to patch up the holes, so we had to go to the big end of the island to find that dark heavy clay to patch the holes. When we were sent up there to load the trucks with that heavy clay, the fighting was still going on.

"One day Oswald Newland, a fellow from Oregon, and myself went to work in the clay pit. The clay was used to build the roads too, including the road up Suribaci.

Newland drove the shovel, a big machine on tracks. It was slow going; that shovel just crawled along. I walked, I was his guard. We worked twelve hour shifts. Trucks came all day, then all of a sudden, they stopped coming.

Night came, and Newland and I were stuck up there all by ourselves with no transportation and no way of knowing what had happened or what was going on. I spent the night by a rock with my carbine at the ready and Newland holed up in the cab of the shovel. We were in enemy territory without any back-up. That was a long night. We sure were glad to see those trucks pull up the next mornin'! Come to find out, the Japs had become active and the officers ordered the trucks to stop in order to protect the drivers. The Japanese were shooting at our trucks, but they didn't realize that our steerin' wheels were on the left side, I guess, and they were shooting at the right side of the cabs…lucky for our guys."

The sunlight coming through the barn door had moved onto his face so he plucked his John Deere cap off his knee, put it back on to shade his eyes, then continued. " I knew a lot of the guys who worked on the road up Suribaci: our commander Ermillio, ole DeRamus, Peterson, Gearheard, and a few others, but we were so busy that it wasn't until later that we realized we had passed each other, probably numerous times on the island without knowing it. When we compared stories, we realized we were often experiencing the same events, just from different vantage points. It was, after all, a small island.

"If the Seabees didn't have the equipment we needed, our guys just built it on site. We had a rock crusher they built that we used to make the gravel for patching the runway and building the roads. It was amazing what some of those engineers and mechanics could do. That was where I first saw R.G. Letourneau's invention, what we called a Tourna-pull. It had big tires and unlike a regular bulldozer, it could turn quickly and it helped us make quick progress on the airstrip repairs and building the roads. He was a genius…lived out his later years in Longview, Texas."

"Not to change the subject," I said, "but how long did the Marines stay on Iwo?" Dad thought a minute, and then said. "As far as I know, the Marines never put up any tents. They finally put up a medic tent, for first aid, on the beach. The Marines started leaving the island after six weeks; when they'd declared the island 'secure,' but there was still plenty of Japanese in those holes to give us trouble.

"In fact, after one incident, they had to call some of the Marines off the ships that were all ready to go home, back onto the island. This was one of the things that families back home probably never heard about…I don't recall hearing anything ever being said about it, but there was a squad of about forty flyboys, all highly trained fighter pilots that came onto Iwo one day. They set up their tents on a ridge above our foxholes. We wondered about the wisdom of settin' up tents when we were still under restrictions, but we weren't in charge. That night, the Japanese organized and came out of the caves and pulled a 'banzai.' There were about 200-300 Japanese involved in the surprise attack. They ripped into the tents killing all the pilots; at least as far as I know, none survived.

"We were under strict orders to stay in our foxholes. We heard the commotion when the Marines opened fire; I imagine the Marines were shootin' at anything that moved...that's about all you can do at night. They killed all of the Japanese attackers, but I don't know how many of our own men might have been killed by 'friendly fire.' They call it 'friendly fire,' but they are dead just the same. I doubt their families ever knew what happened."

He slowly shook his head. "It was chaos up there. It was night, so you really couldn't tell what was going on. I always wondered why they didn't set up a watch, guards posted by those tents, somebody to alarm them. They couldn't have stopped the Japs from killing, but I think they could have stopped them from killing so many. That really shouldn't have happened. But I guess that's just part of war...things happen."

Dad took his cap off, absentmindedly picking bits of hay from the brim. I waited, thinking about those young men lost that night on Iwo, and how that must have been a living nightmare, but unlike mine, one they never woke up from.

Clearing his throat and placing his cap back on his knee, he continued. "About a week or so into the battle, the big brass had already notified Washington that the island was 'secure,' but it was anything but that. I don't think many people realized that until much later."

I agreed. "I've talked to people who thought the battle was over when the flag went up. They were shocked to learn that it was just beginning."

"Yeah, I think a lot of folks have that notion," he agreed. "Lots of misinformation about Iwo. One time in Washington, D. C., at one of our get-togethers, a fella argued with me that there wasn't a ship named the *Bismarck* in WW II. He said it in front of a lot of folks; it embarrassed me, but I got the last laugh. When our main speaker, a high ranking Navy officer got up to speak, he mentioned the aircraft carrier, the *Bismarck,* getting sunk.

"Later that evening, that fella who had argued with me got on the same elevator with me and some of my buddies. He took one look at me and said, "You do well!" It was pretty amusing. The reason I knew personally about the *Bismarck* sinkin' is that some of the sailors who survived were brought ashore on Iwo." He smiled. "That fella thought he knew what he was talking about, but he sure didn't.

"The 58th Task Force - the *Bismarck* was one of their ships - the Task Force lay between us and Japan, protecting us from air attacks. As long as they were there, they shot down everything that headed towards us; nothing got through except once, a lone reconnaissance plane, but it was flying too high for our guns to reach." He chuckled softly. "They shot at it anyway. I remember the ships off shore started shooting at it and we thought it was the Japanese plane shooting at us; we could see the tracers, so we kept our heads down till we found out it was the ships shooting at that one lone plane...wasting all that ammunition."

He seemed to get lost in his thoughts for a minute, and then frowned slightly before continuing.

"When we were there under fire, you could hear those incoming shells whistlin' and you could tell by the sound about how close they were. But you didn't know when one would have your number on it. You were living on the edge, but there was nothing you could do. Finally you get kind of numb to it and you say, Well...I'm either going to live or die, so you let it go. You just do what you're supposed to do." He shrugged his shoulders.

"I need to switch channels," he said abruptly, his unique expression for moving on to a less dire or personal topic. Raising his right arm into the air, he motioned with his hand palm up, indicating a change of direction in the conversation.

"On another note, we were workin' long hours, 11-12 hour shifts. I was driving a truck and we had to service it for the next shift when we came in. Some of the guys driving bulldozers and other heavy equipment started developing back problems because of the long hours and no breaks. We were under pressure to get that big airstrip, the Number Two, finished. About the time we got it completely finished and got the fuel system set up, the war ended. All that work...left to the Japanese." He shook his head, then leaned forward and put his elbows on his knees, staring at the sawdust-strewn concrete floor. I wonder if he was seeing instead a finished, unused airstrip halfway around the world on a tiny island in the Pacific.

I stood and stretched, still feeling the effects of my tense night. I picked up my bucket and put it back into the corner of the barn.

"Going to fix some stew today...do you want any cornbread?" I asked, as much to dispel my still somewhat somber mood as to get information.

"Sounds good to me," he replied, coming back from wherever his thoughts had taken him. "Think I'll work on getting the lumber together for the troughs. You can tell your Mama I'll be home in a little while."

"Ok, I will." I turned to head out the big sliding door. "See you later."

He nodded, got to his feet, put on his cap, and turned back to his work.

On the way home, I wondered if he gets tired of me asking all these questions; I wouldn't want to be a pest. Then I remember him telling Mom one time when I was little, 'She asks so many questions because that's how she learns.' I guess Dad's right…I'm still learning.

I can just see the corner of the Ewing's tin roof peeking out from the sycamore tree. In the sunlight, it seems to sparkle as the southern breeze moves the tiny spring leaves back and forth.

Everything looks so familiar, yet it all feels new...a strange feeling, the same one I'd felt earlier in town. Belonging, yet not belonging, my heart and my head can't seem to make a decision.

One thing I do know, I am ready to see Mama. That was one of the things that kept me going all those long nights on Iwo... thoughts of home and coming back to Mama. Now that I am walking towards home, it is almost too good to believe.

As I pass in front of the Ewing's house, I see Mrs. Ewing out in the backyard hanging out clothes. She is a heavy woman wearing a drab gray housedress and faded apron, but today, she looks beautiful. I don't say a word as I walk by and she doesn't notice me. I really don't want anything to break the spell. It feels like a dream and I guess I am still afraid of waking up.

A little further on I catch a whiff of the wisteria that always covers the old Skeeter place. In the wintertime that place looks haunted, all covered with twisty vines and dead branches, but in the spring, it is covered in glorious purple and white flowers that perfume the entire country lane.

Realizing I had been walking faster than I intended, I slow my stride again, soaking in every detail as I walk the last one hundred yards to home.

Chapter Eight

Winter passed swiftly and now the earth was waking up from its short sleep. In Texas, we joke that we have two seasons, summer and winter. It's a rare day when the temperature is just right and you can enjoy being outside without either shivering or sweating. Today was one of those days.

Driving to Mom's and Dad's with the windows down, I took in the sweet smell of freshly plowed earth from my neighbor's garden and the bluebird song drifting on the cool light spring breeze.

I parked in the old orchard spot, now home to one aged pear tree, and headed for the house. I was hoping Dad was in the mood for a drive around the pasture. He met me at the door wearing his usual attire – worn khakis, a long-sleeved green plaid shirt that had seen better days, and his scuffed brown brogans.

"This is a fine day for getting outside!" he said, and smiled as he stepped out into the sunshine.

"It sure is. Want to take a turn around the pasture and check on the new calves?"

"Sure, then maybe we can clean up around the fig trees before those old honeysuckle vines take over."

We headed for Old Blue, his farm truck, then poked along in first gear over the familiar hills and through the newly leafed woods, our conversation mostly about the cows, when we were going to market, and predictions of the weather for the coming weeks. We counted the six newest calves and commented on how much they had grown in the last couple of months and which cows looked like they'd be next to calve.

Dad made a big circle around the place, behind the pole barn, down by the creek and then to the highest point on the farm, one of his favorite places to enjoy the view; from here you can see dozens of different trees around the ponds, along the creek and fences, all sporting a different shade of spring green.

As we pulled up onto the top of the knob, Dad put the truck in park and killed the motor. Suddenly, all the sounds of early spring rushed in through our open windows: mockingbirds, cardinals, and a pair of hawks screeching as they circled high above us. While we were stopped, a low-flying plane buzzed us and dipped his wings in a friendly salute. That must have set him to thinking about the war, because he started talking about some things that happened at the airstrips on Iwo.

"We had got pretty comfortable with the 58th Task Force, all those Navy Fighters and aircraft carriers, stopping everything that came our way. But when they moved them for the invasion of Okinawa, those Japanese bombers, Black Bettys we called them, started coming in. They were just a two motor bomber, but they were the longest range bombers they had.

"We had our radar set up on Mt. Suribaci, but the Bettys would come in low over the water, under our radar, and then…" he motioned upwards and out with his hand. "…they would go up and over the island dropping their bombs. We had some 'long Toms,' we called them, that could shoot about nine to eleven miles and they were radar-operated, but when the Bettys flew under the radar, the Toms wouldn't pick them up.

"One time, I remember, the Japanese flew up and over the island dropping anti-personnel bombs. They were pretty little things, all painted up, and they had a safety device in the nose of the bomb that worked like a propeller. As the bomb fell, the propeller would turn, unscrew and fall out, and then the bomb would explode on impact. The Black Bettys weren't flying high enough for the little propellers to have time to unscrew, so they didn't explode. We had a detail of guys to go around and just pick them up.

"One night some of those Bettys came in dropping their bombs and we had to take cover. The only real cover we had close by was a bluff, a drop-off of about fifteen feet behind our tents. My tent was right next to the bluff.

"All of us guys knew where it was, so we took cover, but some of the guys further out had to make a run for it. We figured everybody knew where that bluff was and how to get there, but there was a group running for it that night and one of them didn't know where it was, or forgot."

Dad bent over the steering wheel and started laughing.

"The moon was shining bright that night and I watched this short stocky fella go out over the edge, his feet just a pumpin' but there was nothing but air under him."

We were both laughing by that time. Dad threw his head back making bicycling motions with his arms.

"That poor fella squalled out till he hit the ground. It was pretty comical."

When I finally managed to stop laughing at the cartoon-like image in my mind, I started watching a flock of white cattle egrets that flew up over the tree line and banked over the herd of cows below us…it was a pretty sight. The birds slowly settled among the grazing cows and started feeding on the grasshoppers and other insects the cattle stirred up as they moved slowly across the pasture. A few doves were startled by the newcomers and took flight, disappearing over the pond dam nearest us. We sat there in silence for a while just watching, then Dad broke in.

"Not a lot to laugh about then," he said softly. "I remember the night that Meeker was killed. I wasn't workin' that shift, but I remember that Clifford Paxton had the day shift. He was driving the truck to the rock crusher and haulin' the gravel to the airstrip, and Meeker had the same duty on night shift. George Young told me what happened that night…he was there. You remember George," He glanced at me.

"Yes, I do." I said nodding.

"George was one of the guys on the boxing team. He was from Texas too. George and Meeker had been together from the beginning in Bermuda, to Endicott, and California…all the way to Iwo.

"Anyway, George always said, 'They call Iwo *Hell's Acre* and grant you, every inch of it was hell on earth.' That night, George and Meeker were haulin' gravel from one end of the island to the other and had dumped their load of gravel around 4:00 a.m. They were driving beside the road along the airstrip when they had to stop and put out the truck lights because a plane was coming in to land. There could be no lights on except the field lights so the pilots could plainly see them. It was a bad mornin'…cloudy with light rain…and smoky too, George told me.

"He said he pulled his truck in behind a light tower- they were about twenty-five feet high - and his buddy Meeker pulled his truck in behind his, then a Jeep pulled to a stop there too. George said they were just standing by Meeker's truck talking when he saw the plane circle to land and start downward. It looked as if the plane was missing the strip, but George said they kept talking and didn't give it much thought. He said they both figured the pilot would pull over and make his landing okay since he was still a good distance away. George said the next few minutes happened so fast…George said Meeker got a funny look on his face and dashed around his truck, and George turned around in time to see the B-29 knockin' over a light tower, not twenty yards away. The plane was headed straight for George and he started to run from in front of it, but realized he could never get away from the wingspread. About that time, he said he felt something or someone hit him from behind and he hit the deck…said it felt like he had been shoved down.

"The plane hit the tower, the Jeep, George's truck, and was tearing Meeker's truck apart when it rolled over George. The body of the plane went over him and kept going. He said he was in shock that he had escaped. Then, according to George, his real scare came…he jumped up and called out to Meeker. No answer. He said he grabbed a light from somebody who had run up to see what had happened and told them to get an ambulance while he ran to find his buddy.

"He said he found him about fifteen yards from where he had last seen him…he was dead. The only thing George said he could figure was the wing of the plane dipped while hittin' the trucks and struck him.

"Poor George could hardly talk by then. I felt real bad for him. He told me, 'I loved him like a brother. I still can't understand why it wasn't me instead of him.'"

Dad looked out his driver's side window. "I guess a lot of us felt that way too. I don't think George ever forgave himself for not taking care of his buddy." He continued to stare off into the distance and then sighed. I wondered if he was seeing a rocky smoke-hazed island in the Pacific instead of the peaceful landscape around us. He didn't say anything for a few minutes, then glanced towards the house and said, "Guess we ought to be gettin' back to check on your mama."

I nodded, thinking about George and Meeker and how the scars left by war never completely heal.

He cranked up Old Blue, shifted into low gear and we eased down off the top of the knob towards the house. As we bounced slowly along, he told another story about Iwo, maybe to clear his mind about Meeker.

"We had an ammunition dump, we called it, on Iwo. We made the mistake of piling all our ammo in one spot and the Japs were able to hit it. It was quite the fireworks. One of our guys was in the vicinity and started running when it blew. Since it was night and they couldn't see good, they thought he was the one who blew it up and they opened fire. Ended up shooting him several times, but he lived.

"Another time, a submarine spotted a big bunch of tar that we used for makin' asphalt for our roads and airstrips. It was in 55 gallon drums. We had thousands of drums of that tar stacked on the north end of the island, right out in the open. Apparently, the Japanese submarine thought it was our fuel supply and they bombed it. All it did was cause a few fires and we put those out pretty quickly. Our fuel supply was over on the west side of the island in rough territory. It was camouflaged to where it looked like the rest of the terrain and they never did bother it. We got a laugh out of that."

Dad grinned and slowed down as he approached the gate. I hopped out and opened it so he could drive through and park under the shed. After he parked, he took the time to roll up the windows, since the weatherman had predicted spring showers for later. Then he got out and walked to the patio where I was scratching Hobo, their black and white tuxedo cat, behind the ears.

Dad put his hands on his hips and looked towards the fig trees on the west side of the house. "Well, guess we'd better tackle those vines. Hate to waste a pretty day and not get something worthwhile accomplished."

"Ok, I'll go get the hoe and the clippers," I announced heading towards the shop with Hobo hot on my heels. That cat acts more like a dog than a cat, always wanting to be where the action is.

"Meet you out there after I check in with The Lady," Dad said as he walks towards the house.

I smiled. That's what he calls Mom, especially when she is in one of her "I want to be waited on" moods. After over seventy years of togetherness, they still manage to make marriage look easy - no small task.

By the time Dad came back outside, I had gathered the clippers and machete we needed to clear out the tangle around the fig trees. I noticed he had exchanged his brogans for his tall black mud boots. "What did you put them on for?" I asked, pointing to his boot-clad feet.

"They're good for wadin' through briars and vines too," he said, as he patted his stained worn khakis. "Besides," he said grinning, "I wouldn't want to tear up my good britches!"

Raising my eyebrows, I said, "I think you're determined to make those pants last forever."

"Well," he said, patting his leg again, "We've been through a lot together…kind of like these old boots…they've taken me through some tough places - briar patches, mud holes, cattle lots, and flooding creeks. I just can't toss them out when they've served me so well."

I shrugged my shoulders, not able to argue with him. Besides, I'm not one to be talking, I thought as I looked down at my own worn buffalo-checked flannel shirt.

He and I spent the rest of the afternoon getting the fig trees in decent order again, yanking honeysuckle vines from around the trunk and limbs, clipping a couple of young oaks that had decided to make the fig tree their new home, and pulling up some weeds sprouting underneath the spreading branches. All of this activity was conducted under the watchful eyes of Hobo – who trotted out with us and flopped down in the shade to watch.

That evening, after I'd fixed a bite for them to eat, he talked a little more about Iwo.

"We finally got the Number One airstrip finished, but it was short, only about 6,400 feet long. B-59s could land on it, if they hit it right at the front end and braked as quick as they could. One time, a plane landed and didn't brake soon enough and it went off the end of the airstrip."

Dad shook his head. "It's amazing what those pilots can do. We had just finished the Number One when a B-29 tried to come in with its rudder all shot up. They waved it off twice and then, the third time, they gave him the landing signal and he set the plane down. He went pretty straight until about a third of the way down the airstrip, but he was still going about a hundred miles an hour. Then all of a sudden…" He waved his hand to the left, "…it turned and came right towards us.

"We were workin' on the apron at the time. We had a maintainer and roller-packin' machines on the apron. The plane hit the maintainer and ripped the fuel tank open and it caught fire. The operator jumped off, but he was burned. The wing went right over my head and the rest of the work crew. We ran for our lives! The fuselage broke apart behind the wings and the crew poured out. They all made it to safety.

"Then the plane plowed into a cluster of P-51s, armed and ready for action. Several of them burst into flames. The crews managed to get most of the P-51s back away from the fire, all except five - they burned up. As the planes burned, the ammo went to poppin'. We got behind some sandbags. Then the rockets started exploding and those sand bags wouldn't do, so we slithered off and slid down an embankment about a hundred yards away."

"Mercy!" I said. "That must have been scary."

Dad nodded. "We moved pretty fast. There was no time to think. We just ran and dove for cover. It could have been worse. Thankfully, the plane's crew was safe."

By this time, the late evening sunlight was striking the tops of the trees and hills, making them glow with that early vibrant spring green color that only happens for a few weeks in early April. It was one of those moments that made you want to linger: Dad must have felt the same way because he followed me out to the patio talking about plans for tomorrow.

It was a good way to end the day, standing in the gathering twilight, watching the sun dipping lower behind the pines west of the house. He cocked his head to one side and listened for a minute.

"Did you hear that?" he asked.

"I sure did…reckon that's the first one?"

"First whippoorwill I've heard this year…always love to hear them holler."

He stood there straight as an arrow, thumbs hooked in his front pockets, head slightly bent, listening to the night sounds.

"Sure like hearin' them play their song," he said.

I agreed. "My favorite spring sound."

We said our goodnights and as I drove away, Dad raised one hand in farewell, his tall frame silhouetted against the rays of the setting sun, his hair highlighted in the evening's last glow.

The muscadines are ripe. Me, Floyd and Monroe are itching to get through with our morning chores so we can hit the woods. Our old milk cow is being especially stubborn today and has already nearly knocked over the bucket twice.

Usually, I am patient with Old Bessie, but today I'm already hot and sweaty and I have no more patience. The August heat will do that to you – make you short-tempered.

I see Floyd and Monroe pass by totin' Mama's firewood to the back steps near the kitchen and then take an armload over towards the big blackjack oak at the edge of the yard.

I can't see her but I know Mama has been there since sun-up filling her two big black wash pots with water and building a fire, getting ready for wash day. That's why I'm milking this morning. Mama usually does the milking, Old Bessie likes her, but today's wash day and Mama doesn't have time.

Even in the shade, it's steamy. I feel sorry for Mama having to stand over those boiling pots all day, just so we'll have clean clothes.

Listening to the rhythmic sound of the milk hitting the bucket, I let my mind wander back to muscadines...juicy purple-skinned treats just waiting to be picked. We eat them till the corners of our mouths crack from the sweet tart green insides. Mama likes to use them to make jelly, if she can get us boys to pick enough without eating them all before we get home.

Bessie shifts her weight again and smacks me across the face with her cocklebur infested switch, causing me to nearly kick the bucket of milk over myself.

I push against her fat tan side until she steps away from me and I grab the bucket of milk and set it behind me. Then, I untie Bessie from the post and lead her back to the pasture and turn her out. She saunters off with a final swish of her tail and I am glad that chore is done.

When I get back to the bucket, I pick it up and head to the house to get a clean gallon jug. Pouring the creamy warm milk into the jar, I find a lid and ring and tighten it as hard as I can and head for the well. The bucket is sitting upside down by the thin rope we use to tie it off. After carefully attaching the rope to the bail on the bucket, I place the glass jar of fresh milk inside. The rope is already tied off to a short stump of an elm nearby, so I carefully lower the bucket down into the well, just above the water and to the side, so Mama can still draw water without the milk bucket getting in the way.

Come supper time, that milk will be ice cold, a real treat on a hot day like today. That done, I walk around the house to ask Mama if there is anything else she needs me to do. Mama looks up. A strand of hair has worked loose from her neat twirled braid and falls in a brown curl across her cheek. She smiles, sweat dripping off her nose and chin, from bending near the already simmering wash pots.

I watch as she lifts the punching stick and moves a load from the wash pot with lye soap to the clean rinsing water of the second pot. I see that a couple of our shirts are already hanging on the yard fence to dry.

The thought of leaving her with all that work and running through the woods for muscadines makes me hesitate. I guess Mama can sense the struggle I'm having because she says, "You go on now and find those muscadines…it's supposed to be a good year."

She smiles again and shoos me off with a wave of her hand.

Floyd and Monroe are already headed down the path to the woods across the road, but I stand rooted, not able to decide.

Seems like Mama can always read my mind…I see her pick up the empty bucket she used to carry water and she holds it out to me.

"Go on…you'd better pick me some before they eat them all." She grinned.

I grin back and reach for the bucket as I run by.

Today, I enjoy the muscadines more than usual.

Thinking about Mama and all she does to take care of us… makes me put more muscadines in the bucket than in my belly. Maybe this year, Mama will be able to put up more jars of muscadine jelly than she's ever put up in a season. And I bet it will be the best-tasting too.

Chapter Nine

"Looks like this will be the last time we'll have to hay this season," Dad said as we headed to the barn to get the tractor.

"Good thing we got a few extra bales from that second cutting, or we wouldn't have made it," I replied as I grabbed my gloves.

He nodded. "Always like to have a cushion; you never know how cold a winter will be."

When we got to the shed, Dad climbed onto the big John Deere while I pulled Old Blue around to the barn and loaded up a couple sacks of range cubes. Then I trotted around the corral to open the gate to the pasture. It is getting harder and harder for him to get off and on the tractor, so we try to save him all the steps and ups and downs we can.

This has always been one of my favorite tasks on the farm. We used to put out square bales, which was a lot more work, but we didn't mind. Any time spent with Dad was a plus in our book, and we kids never missed an opportunity to help out when we could. Watching him load the heavy round bales on the tractor forks and ferry them out to the hay rings, reminded me how blessed we've been to have him still around and able to do the things he loves.

After we put the rings around the bales and dumped little piles of cubes here and there on clean spots of sprouting new grass, we got back in Old Blue and headed home. He drove extra slow from the barn, enjoying the view across the pasture to the ponds.

"Always loved the springtime," he finally said, then in such a quiet voice that I almost missed it, "One more time." Dad has never been one to take anything for granted. He is thankful for every day and the beauty in it, knowing that tomorrow isn't promised to any of us. Hearing him say "one more time" as he does a difficult chore or sees a new season unfold, never fails to hurt my heart a little because time is more precious with every passing day. Dad has always appreciated the blessings he enjoys and in turn, I believe it has made us more thankful too.

I wondered if he would be up to reminiscing. I had a few questions since our drive last week, so when we pulled up to the shop, I asked. "Do you think you could tell me about the last bit of time you were on Iwo? I know you were there till right after Japan surrendered, but I want to know what you did on Iwo those last few months."

He turned off the key, rested his right hand on his leg, and stared straight ahead for a minute. "I'll need a cup of coffee to jog my memory," he announced, then grinned at me.

Laughing, I followed him into the house and watched as he went through his routine of fixing a pot of coffee: eight cups of water and three-and-a -half scoops of Community Breakfast Blend coffee. In just a few minutes, we were out on the patio sipping the delicious smelling brew. It didn't take long for Dad's memory to kick in...with the caffeine's help.

"Lots of things were happening that last few months we were on Iwo. Of course we had no idea that Japan would surrender, so we were preparing for an invasion...at least that's what we figured. One of the things I remember after we finally got settled into our housing is we started to get mail from home. It was like a life-line. I read and re-read my first letter. It seemed like a lifetime ago since I'd left home.

"I had a surprise visitor while I was on Iwo. An ole boy I knew from back home - J.C. Loper came walking into my tent one day. I sure was glad to see him. He had asked his aunt, Mrs. Skeeters, our neighbor, for my address and how to find me. He was in the 133rd CBs, I think; he landed with the 4th Marine Division. We were both driving trucks on Iwo and had probably passed each other every day, but that was the only time I saw him while I was there."

"There were also some famous visitors while we were on Iwo; Gene Autry came and sang. I heard from some of the guys that he sang on a platform on the beach and would sing any song they would call out. That was pretty gutsy because at that time we were still getting Japanese sniper fire. Buddy Griffith came to the island too and brought some of the Major League baseball players. Some of our guys got to shake their hands and talk to them. I remember seeing Joe E. Brown. He put on a show… but back then, not much was funny to me.

I also remember one show that had a young lady who was an acrobat. She got up on this little stool, about a foot high and put a fifty cent piece between her feet." He motioned in the air as he described what happened next. "Then, she bent over backwards, all the way and picked up that fifty cent piece with her mouth. Then, she caught her ankles and walked off the stage lookin' back between her legs…it was spooky. Looked like a spider! I didn't see how a human being could do that.

"We also had some singers come, but I don't remember who they were.

"Earlier, they had put up a movie screen and would show movies at night. They had piled some sandbags for us to sit on to watch the show. They caught one of the Japanese sneaking out of a nearby hole and he said he had been coming out and watching the movies. We were still in our foxholes at that point. I guess he'd had enough of stayin' in that hole. I don't blame him; we were all pretty tired of those holes.

"You see, the war wasn't over…when we got our job done on Iwo Jima, we were fixing to invade Japan. Things were lookin' pretty serious for us until they dropped those super bombs and ended the war. If they hadn't, there is no telling how long they would have fought and how many more lives would have been lost. The ones runnin' the war, Tojo and the others, their necks were on the line and they had nothing to lose, so they would have kept on fighting. They weren't out there on the front lines anyway.

"There were times when I wondered if all the sacrifices I had seen was worth capturing and holding that volcanic mound of rock. Then one day we had a huge storm and all of our B-29s had to land on our airstrip. I realized at that point that if we didn't have Iwo, all of those planes and pilots would have been lost."

Dad stared off into the deepening blue of the evening sky, sipped his coffee for a minute, then continued.

"There were many more times that planes had to land to survive, either disabled, shot by enemy fire, or running out of fuel because of a long-distance bombing run. That airstrip saved a lot of lives.

"One time a B-29 stopped over for a repair and our Skipper invited the pilot to spend the night. He stayed and ate supper with us, but then he said, 'I believe it's a little too unhealthy here. I believe I'm going to take on off.'" Dad grinned. "He was probably right. Another time, a B-29 came in on Number One airstrip and couldn't stop and ran off the end and crashed.

I don't remember them doing anything with it like salvaging it. We also had a fighter plane, a Thunderbolt I believe, that undershot his landing and slammed into the bluff at the beginning of the runway.

"Sometimes those fighter planes would come in with bullet holes all in them from anti-aircraft guns. I remember overhearing some of the pilots talking about how they'd 'get a feeling that this was it…they weren't going to make it back.' That was tough on them…heading into battle knowing they might be shot out of the sky.

"I was driving a truck along the airstrip when I had another close call. There was a high place towards the end of the Number One airstrip and it sloped down the hill at the end of the runway. We'd have to stop at the top of this hill when a plane was taking off and wait till it cleared. I was waiting at the top of that hill for a P-51 to take off, and as he lifted off, one of those big rockets come loose and it went tumbling down and went under my truck. Thank goodness it didn't explode. But it all happened so fast, I didn't have time to get scared…not till later."

"Oh my word!" I exclaimed, mouth open in amazement. "What did you do?"

He kind of smirked. "I drove off."

We both laughed.

"There was another B-29 I remember that was so shot up it couldn't even attempt a landing. The landing gear wouldn't come down either, so they circled the island. On one pass, the crew jumped out. They made another circle and the pilot and co-pilot jumped out. They had set the automatic pilot so the plane would go out to sea and then they sent a P-61, a night fighter called a Black Widow, to shoot it down. We saw the whole show.

"Another thing…I can't swear by it, I just heard about it…there was a B-29 that for some reason the bomb bay door opened as it went over the island, and it spilled a bunch of boys out. Of course that was the end of them. Now, I don't know that for a fact, but that is what I heard happened.

"Below where we'd been dug in, on the beach, a B-29 made a water landing because the landing gear wouldn't come down. That was pretty risky.

"We had a P-61 or Black Widow come in at night and undershot the Number One airstrip…that was the end of them. I think they had a three man crew, the pilot, a radar man, and a gunner. Black Widows had about four 20 mm guns in the nose and those things were deadly. They could really knock stuff out of the sky. They had a twin fuselage with the cockpit in the middle. It was a pretty heavy plane and they were landing probably a hundred miles an hour. They didn't have a chance...there was so much loss and devastation of equipment and men until you just had to kind of close your mind against all of that and do what you were told to do."

Dad's eyebrows knitted up in a pained expression, then he looked away before continuing.

"Later on, we finished the second airstrip; it was 9,600 feet long. It was big enough to allow several planes room to land."

He punctured the air with his finger as he described the airstrip's capabilities.

"One would be almost at the end, one in the middle, and one coming in to land – all at the same time. We had a big circle they could make where we had fueling stations and they could be refueled and off again. They wouldn't be on the island long that way.

"One thing about the island a lot of people don't know, it was rocky, but on the big end of the island was this hard red clay. Over eons of time the elements had carved it into deep ravines and steep cliffs, like the cliff close to our chow hall. This was where one of our 31st guys, a fella named W.T. Rich, carved a sculpture of the flag-raisin' on Suribaci. His nickname was "Art." He was from Inglewood, California. Old Art did a good job. The sculpture was preserved and you can still see it on the island today, if you take a notion to go there.

"One thing about those ravines, they could be dangerous...downright deadly. The Japanese would set bombs and booby traps in the ravines. When our tanks tried to go through them, they wouldn't make it. In one place I saw four tanks, blowed inside out...all in a row. The weakness on those tanks was the bottom. When they ran over a land mine - that was usually it. The gun turrets on top were real heavy and on those four tanks I saw, the turrets had blown up in the air and come down on top of the tanks, upside down. Those guys were gone instantly...they never knew what hit 'em."

I winced at the thought and shivered at the gathering chill of the evening which felt sharper than usual. I glanced over at Dad and wondered if he didn't feel the same chill as he put his hand to his collar to pull it up a little against the cool air. He wrapped his hands around his cup as if to warm them and took another sip.

"The P-61s and our shore guns gave us good protection. We never had any effective Japanese bombing of the island while we were there. The P-61s patrolled at night and stopped most of those Black Bettys trying to reach the island. Those anti-aircraft guns could reach nine to eleven miles and combined with the Black Widows, they kept most everything from getting through. We only had a few scares where they managed to get through by going under our radar. Like when they dropped those anti-personnel bombs that didn't explode.

"The Japanese also had what we called "nee mortars." These didn't have any fins to guide them; they would just tumble. A lot of them didn't go off, but if our mortars hit nose first, they would explode. The Japanese had planes called Zeroes - they were like a flying tin can and could cut circles around our P-51 fighter planes. The P-51s were heavy, but the Zeroes were real light. I think that's why they called them kamikaze pilots – if a Zero was ever hit that was it. I imagine the pilots knew that and they crashed into ships or whatever to cause as much damage as they could. They knew they were done for; most of them probably went into it knowing they wouldn't survive."

"Speaking of not surviving, whatever happened to Whitey Knerr?" I asked. "I know he died on Iwo, but how did it happen?"

Dad thought a minute then answered slowly.

"Every time I think of Whitey, I think about when we were leaving the States. His words as we pulled away from the California coast were prophetic. He said, 'I sure hate to leave the States. I don't think I'll come back' and he didn't. When we got to Iwo, only our outfit - C Company, and a few others were called in early on to help clear the beaches and such, but Whitey wanted a part of the action, I suppose, so him and another fella jumped ship. They weren't supposed to come ashore yet, but they did. Whitey disappeared and the other fella didn't come clean about it for two months. He finally told us that he and Whitey went into some caves lookin' around and came upon some Japanese. They started firing. The other fella turned and ran, but not Whitey. He dropped down on one knee, according to the guy that was with him, and the last he saw of Whitey was him returning fire with his carbine. He told them where they would probably find him and sure enough, they found his skeleton in one of the caves. They knew it was him because his dog tags were still on his skeleton."

"Why do you think he went poking around in the caves?"

"Probably souvenir hunting...maybe he wanted to use some of his training...I don't know, but I sure hated it happened. Whitey was a nice person."

"Sure is a shame...can't believe that other fellow waited so long to let someone know what happened. Sounds like a coward to me," I frowned.

"Well…it's hard to know what you might do in a situation unless you're there. Guess he was ashamed of leaving Whitey to defend himself. Maybe that's what took him so long to confess."

We both sat in silence for a while watching the evening shadows march across the hills and valleys of the farm. A fresh earthy fragrance wafted to us on the light breeze and the last trill of a mockingbird sounded nearby. Soon, I'd have to head home, but not before I finally asked Dad a question I had been wondering about for a long time.

"I know it's hard to think about," I hesitated. "but, how did y'all bury all those poor men on Iwo?"

Dad didn't answer right away. I could tell he was pondering his reply by the way he sat back in his chair, staring straight ahead. He drank another sip or two from his cup, then set it down on a low cedar block "table" by his chair.

"The 31st buried the 5th Marine Division dead, and our own 31st dead. They surveyed off the area, laying it out symmetrically and using big machines to dig the trenches. It was a long flat place, just so deep." He measured in the air with his hands. "They laid out the dead in order. You had to have a good bit of time to bury all those men. Nothing was done haphazardly. I didn't have a part in burying the dead, and I'm glad about that. It was a tough job mentally.

"After they laid out the bodies, they left one dog tag on the body and put one dog tag on the cross above the body. The Jews had a Star of David on theirs. It was laid out where we knew where every man was…that was the importance of dog tags."

Dad's expression clouded. He was quiet for several minutes then finally said, "I knew several of the guys that were on that detail…it took its toll."

I imagined row after row of white crosses. It was a sobering thought. Images I had seen of Arlington Cemetery came to mind. Except these brave men were all buried on a forsaken island in the middle of the ocean with few, if any, visitors to their grave. It seemed so lonely and sad.

"Who painted the crosses? I know from the pictures in your Seabee book that they were white."

"Martin Devries, and I believe, a fellow named Strictfield were the ones who painted the crosses. It was kind of ironic that when Martin got home he eventually became a commercial paint contractor in Houston. He was pretty successful, doing big jobs in the city. We enjoyed getting to know Martin and his family at some of our 31st get-togethers. He always said that painting those crosses on Iwo Jima was the hardest job he ever did."

Dad picked up his cup and sipped the last of his coffee as we both sat there in the quiet of the evening. As dusk settled around us, I felt a change in the wind and a heavier chill in the air and couldn't help but smile.

"Feels like winter might not be through yet," I said.

He set his cup down again and raised his hand to test the air.

"Feels like it's shifting around to the north again. We might get to have another fire or two before we shut the fireplace up."

"I hope so." I stood up and stretched. "We'll be wishing for these cooler evenings come July and August."

"That's for sure." He caught both arms of his chair, slowly stood, and straightened.

"Guess I'd better be getting home and feed my critters," I said as I gathered our cups.

Dad stood with his hands in his back pockets looking off towards the knob. "Think I'll just enjoy the cool for a few more minutes."

"Ok, see you tomorrow. Love you." I went through the back doorway to leave our cups at the kitchen sink and say goodbye to Mom.

"Love you too," he answered as the door closed behind me.

When I came back out, I saw his silhouette against the gathering darkness. Not wanting to disturb his thoughts, I slipped past him, then eased into the Jeep and headed to the next hill and home - already looking forward to our next talk.

The old T-model truck chugs along in the growing twilight as we make our way to Tecula and the tent revival meeting that has been going on all week. I'm excited- we all are - we don't get to go many places, especially at night. Mama warns us about trying to jump out of the truck before we even get there.

Soon I can see it, a big canvas tent top set up in a clearing next to the railroad tracks. After we get closer, I can see they have wood shavings on the ground all under the tent. I guess so the 'Holy Rollers,' as my dad calls them, won't hurt themselves if they fall down. I love the way the shavings smell, all fresh and clean.

Mama fixes us pallets to lie on when we finally make it to an open spot and park behind the tent. The younger kids usually don't last through the whole meeting before they fall asleep, but I am almost six, so I am determined to stay awake. Last night, I overheard Mama saying that the meeting would probably go on till after ten o'clock.

While Mama fixes the pallets, I watch the man lighting the lanterns under the tent. He pulls on a little rope to lower the lantern, lights it with a match until it glows bright, then pushes it back up out of reach. One by one he makes his way around the tent till it is all lit up. Another man is bringing more folding chairs from the back of a wagon and setting up another row along the back of the tent.

People are filling up the tent pretty fast, so Mama gives us one last word then goes to find her a seat. I don't like those old wood chairs; they always pinch me when I try to sit on them, so I am glad Mama makes me stay in the truck.

They like to put us kids out of the way because, Mama says, when people start to get the Holy Ghost, they can run over us.

Maxey Davis comes running by and I ask him if he wants to watch with us. He climbs up into the bed of the truck and we get ready for the show.

It is something different, a real treat, to be out at night with other kids, seeing people jumping around and hollering...I can't wait.

In a little while after the singing and some talking, the preacher finally stands up. He starts out kind of slow, but once it is good and dark that preacher gets wound up. He is holding up his big black Bible, quoting scripture, jigging up and down, and hollering for all he is worth about sin and hell and repenting.

All of a sudden, some of the folks get the Holy Ghost. They jump up out of their seats and start tearing around the outside of the tent and back down the middle aisle. Maxey and I look at each other and grin.

Two large women start down opposite sides of the tent, but when they meet going down the middle aisle they jump into the air and slap hands. Maxey and I think that is pretty good jumping for two fat ladies.

We are enjoying the show, but we want the Holy Ghost too. It looks like fun, so we ease down off the truck and start rolling around and jumping up into the air.

When Mama finally sees us, she comes over and puts a stop to our Holy Ghost revival...but boy, did we have a time!

Chapter Ten

We were deep into April and everything had leafed out, even the cautious pecan trees, before Dad and I got a chance to talk about his war experiences again. I was anxious to hear about leaving Iwo and his time in Japan. When we were growing up, Dad told more stories about his time in Japan than he did about the war. He always seemed to enjoy that part of his experience and didn't mind talking about it, except for the devastation at Nagasaki - always the most sobering of his tales. I decided that today, I wanted to hear all the old Japan stories again, hoping that some new memories might surface in the conversation.

I had already made some small crunchy apricot fried pies to take over for coffee time, so I spent a few minutes gathering some necessary items before I left. With my recorder, fresh batteries, pen, notebook, and fried pies in hand, I loaded the Jeep and waved to my husband as he made a sweep across the yard, testing his mower for the first cutting of the year.

On the short drive to Mom's and Dad's, I admired the myriad shades of green that always paint the hills and hollows of East Texas every spring, from the ever-present dark green pine needles to the bright green of new sweetgum and oak leaves, to every possible shade in between. As spring progresses, they all manage to meld into a more uniform-looking landscape, but for now, the variety is almost dizzying.

"Knock! Knock!" I called out as I opened the back door with one hand, balancing all my stuff in the other.

"Come in! Come in!" Dad said as he rose out of his chair at the head of the table just as I stepped through the kitchen doorway.

"Brought some of your favorites…apricot fried pies."

"Woo…wee! Those sure will be good with some coffee," and he immediately made his way across the room to the stove.

"Just what I was thinking." I put everything down on the table and turned to watch him get the coffee pot and fill it with water. "You said this afternoon would be a good time to visit…hope it still is."

"I haven't got anything else to do today, so it's as good a time as any," Dad replied as he assembled his old-fashioned basket and stem and reached for the beat-up tin coffee canister that has been a fixture on the counter for as long as I can remember.

I watched as he measured the three-and-a-half scoops into the basket, pushed the percolator lid down, set it on the back burner of the stove, and turned on the gas.

"How many times do you figure you've done that over the last seventy-plus years?" I asked.

He paused while he watched the flame and adjusted it to suit him.

"Well…I don't rightly know, but it's been a good many times."

Satisfied with the flame height, he headed to his chair where he could keep an eye on the coffee pot. Dad has a habit of getting "side-tracked" and letting the coffee boil over, necessitating a flying leap from his chair to turn the flame down. This never fails to amuse us "kids."

"While we're waiting for the coffee, you want to tell me about your exit from Iwo? You said it was a pretty quick one." I pulled out a chair and took a seat catty-corner from his spot at the head of the oval oak table.

Turning his chair to get a better view of the stove, Dad sat down and said, "Sure, it'll take the coffee a while to perk. This new stove doesn't work as good as the old one, takes forever to get hot enough."

I reached for my notebook and pen and set up the recorder while he monitored the progress of the coffee. I saw him eyeing the platter of fried pies. "Might as well go ahead and try one since we have to wait on the coffee," I encouraged.

He grinned and pulled the platter to him. After lifting the edge of the plastic wrap, he picked a nice brown pie and started munching.

"Ok, I'm ready when you are," I said, smiling at his pleased expression.

He was so intent on that pie that I hated to interrupt, but he leaned back, pie in hand, and got that look that told me he was calling up long ago memories.

"After the bombs were dropped," he said, "and Japan surrendered, things eased up on the island. We had time to relax a bit. We even had time to shave regularly."

He glanced at the stove, still munching on his pie.

"I remember most everybody had let their beards grow because we couldn't shave for a long time. One day, DeRamus, a fellow on my crew, didn't show up for work and I started asking around, Has anybody seen DeRamus? And this ole boy drawled, 'I'm Dee…Raymus.' I looked at his naked face. He had a thin face and a pointy chin, and I told him, Boy… go grow you another beard!"

We both laughed.

"He was one of those fellas who looked better with a beard," Dad added when he stopped laughing.

"I believe it was during that time, after the war was over, that we made our group pictures for our 31st book." He continued. "We had an older fellow as our photographer and when they sent him back to the States, they took all his stuff from him. I guess they were afraid he would let something out they didn't want seen. They even censored the letters we wrote - they all had to pass through the Navy censors.

"Our commander, Ermilio, went home from Iwo, and Commander Church came in to replace him. There was a point system to go home. If you were older, had a wife and children, you got points. My buddies "Coon" Cecil Watkins and "Bud" McEwen both got to go home from Iwo Jima since they were married. A lot of the older guys got to go home pretty quickly, but I was young, single, and didn't have many points, so I was sent on to Japan as part of the Occupational Forces. The brass said we made one of the quickest exits they had ever seen. On the mornin' of October 16, 1945, we mustered and were carried by truck to White Beach where we loaded onto LCTs to take us to the *USS Guilford.* We were among the first to arrive in Japan."

"What did they do with the Japanese that were still on the island?" I wondered out loud just as Dad jumped up to catch the coffee before it perked over. I joked. "You move pretty fast for an old guy."

His expression unchanged, he quipped, "Coffee is pretty important… it inspires motion."

I laughed and watched as he turned the blue flame to low and set the ancient white timer on top of the microwave for the requisite four minutes.

"There…that ought to do it." He walked more slowly back to his chair and sat down. "Now, what was that you wanted to know," he asked, running his fingers through his thick white hair, "about the Japanese?"

"I was wondering what y'all did with the men you captured from those caves?"

"Well, we didn't capture many Jap prisoners on Iwo…most would rather die than be captured. We later learned that they had been told to fight to the death; there would be no surrender. When we did have a prisoner, he was treated humanely. Some of the guys even shared their cigarettes and other rations with them. Unfortunately, **our** guys did not receive that kind of treatment from the Japanese. They were brutal in their treatment of American POWs."

Dad frowned and glanced in the direction of the stove.

"We had a stockade built for prisoners at the foot of Suribaci. Early on, we had only about eighteen or twenty prisoners. The Jap prisoners got to watch the road that they said couldn't be built going up to the top of Suribaci. The Japanese had tried to build a road, but hadn't made much progress. When the Seabees got there, we had that road built to the top of Suribaci in just a few weeks…a road wide enough for two vehicles to pass.

"I guess it's true then," I said. "The fastest way to get something built is to tell a Seabee it can't be done!"

He smiled. "That's part of our motto 'Can Do.' We don't like to be told we can't. There is a saying with the Seabees: We can do the difficult quickly, but the impossible may take a little longer.

"Anyhow, after the fighting had mostly ended on Iwo and the island was relatively secure, they sent some kind of Special Forces American Japanese to the island so they could go into the caves and talk to the Japanese soldiers and tell them what had happened. There were around three hundred prisoners taken after the fighting ended. I don't know what happened to them or where they went after Iwo. We were too busy workin'. An estimated twenty thousand Japanese died on Iwo.

"Did you go into the caves after they got the Japanese out?"

"No, I never went into the caves. I know a lot of the guys did, but I never ventured in any of them. I did pick up a Japanese newspaper, some boxes of matches, and some personal letters I saw on a pile of stuff they had pushed out of the caves. My conscience got to bothering me when I got home…probably after my conversion to Christianity, and I destroyed the letters. I felt like it would be a disservice to them and their families to keep them. They were just soldiers like us, doing what they were trained to do…didn't have a choice probably. But, lookin' back, I could have gotten someone to read them and pass judgement on what I should do with them…but, I did what I thought was best and I don't regret it."

Just then, the timer went off and Dad jumped up to set the coffee pot off the stove. I retrieved two cups from the sideboard next to the table. Mom had decided that she liked them there instead of in the cabinet since they were pretty: white, thin-rimmed china cups with pictures of fruit on them. I got my favorite one with huckleberries and Dad's favorite apple cup and took them over to the counter next to the stove. I poured myself a cup while he fixed his with cream and sugar. I've drunk my coffee black since I was a teenager. Dad says it's because he ruined me on too much creamer when I was little.

After we settled back around the kitchen table and helped ourselves to a fried pie, he continued talking about his time after the surrender.

"We were among the first to arrive in Japan, on the southern island of Kyushu. We sat for two days there at Sasebo before they figured how to get us up to Omura. Some of the Seabees stayed at Sasebo, but most of Company D and part of my company, Company C, went on to Omura, about thirty miles from Sasebo. The bay at Omura was a natural harbor; there was only one way in and one way out."

Dad gestured with his hands as he spoke, leaving a trail of crumbs across the table.

"The Omura Bay was pretty shallow and you had to be escorted by a Japanese guide to get in there, even at high tide. The narrow place we went through was rocky and opened up into a big body of water, surrounded by land." He drew a circle on the tabletop with his finger to illustrate the layout.

"When we finally got unloaded and ready to move, they put us into an old bombed out plane assembly plant that was close to the water where there was a dock. We had bombed that factory all to pieces, but that's where we spent the first two or three days in Omura…sleeping on old parts racks. We later found out what the hold-up was in getting to our regular barracks…they had to disinfect and treat the whole place to get rid of the fleas and lice."

I made a face, "Ewww…that's awful."

He smirked. "It would have been more awful if they hadn't gotten rid of them." We sipped our coffee and polished off the last bites of pie. We sat in comfortable silence for a bit before I asked about their camp in Omura.

"Well," he continued, "it was an old Japanese military headquarters, laid out in a square like ancient towns used to be with buildings all around the outer edge and only one entrance." Dad motioned with his hands as he talked. "When you went in the entrance, there were buildings down the right side, the left side, and in the front, with a big open area in the middle for a drill field. It was several hundred feet square. A road went all the way around the drill field, and the buildings all opened out onto the road. The buildings on the right were two-story barracks and the chow hall was on the left side. On the other end of the field from the entrance were the horse stables, along with more barracks and other rooms. I guess the officers had horses at one time. Of course, there weren't any horses by the time we got there; they had probably already eaten them."

I made a face, horrified at the thought, not only of being at the point of starvation, but also of eating a horse to survive. One of my best friends growing up was my horse, McCloud. I would truly have to be starving to death to entertain such a thought.

"The barracks were unheated and not built like ours at all. They had open rafters - unfinished lookin' to me. When we got there it wasn't all that cold, but later on it did get pretty cold. We had sleeping bags but it was still cold inside those buildings."

His expression lightened at a memory.

"I remember not long after we got to the barracks we were wandering around and saw the Japanese people harvesting turnips, potatoes, and some other kind of big root vegetable. They were still workin' in the fields and stuff when we first got there. Seeing all that harvesting activity made me feel a little less homesick. I liked being around anything that reminded me of home."

"Did y'all buy any of the produce?" I asked.

"No, by then we were well off with supplies and we had our own big galley with walk-in freezers and refrigerators. We had some good cooks too. I remember one big ole boy was from Tennessee. At that time, we were still under quarantine; we couldn't go anywhere on liberty. We had to stay in camp…unless we snuck out." He grinned.

"And…we could get in a heap of trouble for that. We didn't know how the Japanese would react to us taking over their country. As it turned out, I couldn't imagine a bunch of people being as easy to contain as they were – very few episodes of any kind. But on their special holidays, we were cautious not to go out.

"The Japanese were good people. They worked - never created any trouble for the Occupational Forces, at least not that I know of. One Japanese woman told me, 'I'm so glad the war is over, even though we lost.'

"They were prepared for a long war because they had supplies in caves all over the island. I remember seeing one of the caves, big enough to drive a truck into; packed with supplies, just off the road on the way to the rock crusher we were using to repair the roads.

"Some of the guys hated their time in Japan, but me and my buddy Clifford had a pretty good attitude and learned to like the Japanese people we worked with and around. I think that made a difference in how we spent our time there too. It was sad to think that many of them lost family members and friends in the blast at Nagasaki, yet they were good to the occupational forces, at least where I was in Omura.

"One funny I'll tell on myself is about our barber. We had a Japanese barber and I didn't think much about it until he pulled out that razor to shave my neck." Dad laughed. "I wasn't too sure about that."

My eyes widened. "I bet not!"

"We also had a Japanese interpreter at the base in Omura. He invited me to his home to meet his family and have tea. It was a real small house. They didn't have a living room and bedrooms like we do. They slept in what we would call the living room, on mats. They would roll them up and store them in a closet. There was a living area, a kitchen, and a bathroom. The Japanese did something else we don't do; they collected their waste. A person would come around and collect it and spread it on their fields for fertilizer."

I felt my nose crinkle. "Boy that is sure different than over here."

He nodded. "Yeah, it was **real** different. Anyhow, they all spoke English and I enjoyed visiting with them. They had two daughters and one boy. The boy was the youngest. He was about seventeen and the girls were nineteen and twenty. I was a visitor in their home on several occasions.

"One day when I was out driving around, I recognized the two girls walking along the side of the road. I stopped and asked if they needed a lift somewhere, and they said they were going to the hospital to see some relatives. They crawled right up in that big ole 6-by truck and off we went. They told me that nearly everybody in Omura had kinfolk that were killed or damaged in the atomic blast. We drove and drove till I finally asked, 'Just where is this hospital you're going to?' They said, 'Oh...we passed it back there in town.'" He laughed. "They were just enjoying the ride!"

Thinking how funny it was for those girls to just keep riding and riding without saying a word, I got up to refill my cup and get Dad a glass of water.

"What exactly was your job there at Omura?" I asked, as I placed the glass in front of him.

"I spent six months in Omura. My job was keeping up the roads going towards Nagasaki and Sasebo. When most of the older guys with points went home, I was bumped up in rank to second class petty officer and was in charge of the trucks and heavy equipment and maintaining the airstrip and roads. Warrensburg, a Swede, was in charge of transportation and heavy equipment while we were on Iwo Jima, but he had enough points to go home after we had been in Omura for a while.

"At first, me and a fellow that I'd known since Port Hueneme named Seiberlich were assigned as night dispatchers and were in charge of keeping the trucks' flat tires fixed. There was so much shrapnel, blowed up bits of buildings and such all over the roads, that keeping the truck tires fixed was a full time job. Those big old 6-bys, we called them, had double wheels on the back, so they had ten tires on the ground…nearly every truck had flats. We didn't fix them ourselves; we had a crew we supervised to take care of changing the tires.

"After Warrensburg went home, they moved me from night dispatcher to his place. I don't know what happened to Seiberlich, but he must have gone home too, since he was married and had more points."

"I heard you telling Uncle Monroe about some motorcycles you and Seiberlich had. What is the story on that?" I asked, as I reached for another small fried pie.

"There were two motorcycles at the camp and Seiberlich decided he was going to ride one, but ended up runnin' it up one of the support poles at the stables. And some guy got on one and went around on the road by the barracks." Dad motioned in a circle. "He kept going round and round until the Commander had enough of it and took the motorcycles away. I don't know what happened to them, but I never rode them, so it didn't matter much to me.

"Of course, as time went on, we had men being dispatched to other parts of Kyushu on details, so our numbers there at Omura were steadily shrinking. In my new assignment, my job was to keep up the roads halfway to Sasebo and halfway to Nagasaki. Omura was about middle ways between both towns."

I checked the time on my recorder, then asked about his new responsibilities.

"There were three of us second class petty officers: me, Clifford Paxton, and Norman Baker. I worked outside, maintaining the roads, Clifford drove a truck, and Baker worked in the office.

"I was in charge, but I never said much about it. I never mentioned it at any of our reunions, except that time in Washington, D.C. when I mentioned to Baker about him being in charge of the office in Omura, but he claimed he didn't remember anything. I wondered about that...why he didn't remember it. Maybe he felt like **he** was the one in charge, I don't know. But I thought it was strange he didn't remember...but that's his deal I guess."

"How many times did you go down to Nagasaki?"

"I made two different trips down to Nagasaki, one to pick up a huge generator. It weighed several tons. It was a load for that 6-by truck. Later, I made another trip to pick up a piece of road equipment, an asphalt packer. It was only about thirty miles down there, but as the crow flies, it was probably less than half that. The roads to Nagasaki were real crooked and winding, so it took a while to get there. I had all day to make the trip though and that gave me plenty of time to wander around Nagasaki and take a close-up look at the damage. It had only been a few weeks since they dropped the bomb, but nobody said anything about us being around the bombed area. We had no idea we could be in any danger. I walked all over that place."

"Well, doesn't seem like it hurt you."

Dad cut his eyes around at me with a mischievous expression. "Maybe that's why I'm so well-preserved."

We both laughed.

"What did it look like there at the bomb site?

His eyes narrowed and his countenance instantly became clouded.

"When I got there, it was just like the bomb left it, nothing had been touched, no clean-up, nothing. It was total destruction; only a tremendous heat could do what I saw to the buildings, railroad cars, and warehouses - in some places there was hardly anything left at all. The railroad cars were almost melted, just twisted and crumpled up. I saw pieces of china dishes and roofing tiles scattered everywhere. There was only one building left standing, way back off towards the hillside. I believe it was a hospital, but all the windows, doors, everything, was gone, it was just a skeleton of a structure. That whole area was devastated. Most of the effects of the blast were centered on that industrial area."

He traced with his finger on the tabletop as he described the layout of Nagasaki.

"The way Nagasaki sits, it's got the waterfront on one side and the other three sides are surrounded by mountains. Houses were built up on the mountainsides, and all that I saw left of the houses on the industrial side were the bits of tile roof and china pieces. The bomb exploded in the air, so there wasn't a crater like you would expect. Nagasaki was divided, too, by part of the mountain range that came close to the water. There was only enough room between the water and the mountain range for a highway and a railroad, but the other part of Nagasaki, where most of the houses and stuff were, it was still standing. But the industrial part was totally destroyed. I don't know how many it killed on that other side, but as far as I could see, the buildings there were still standing."

"It's sad to think about all those people. Like you said, everyone in Omura either knew someone or had kinfolk there." I shook my head at the horror of war and the inability of man to learn from the past.

"War is a terrible thing…but as long as man is in charge, there will be wars."

I knew he was right; only God can put things in order.

"Can you tell me about the last part of your time in Japan? What was it like?" I asked as I stood to get myself a glass of water, having reached my coffee quota for the day and then some.

He stood slowly and paused to straighten up before gathering his plate and cup and walking towards the sink. After he placed his plate in the sink, he turned and stopped, coffee cup in hand, looking like he was considering his next thought.

"Well, let's see," he said as he moved across the kitchen to reheat his coffee in the microwave. Since he had been doing most of the talking, it had gotten cold.

"Later on, we picked up and moved to another camp area. We had thinned out since a lot of our guys had gone home. There were only about sixty-five of us left; most of those were new guys. We went from the big camp we no longer needed to one we shared with the Marines. By then, we had picked up and cleaned up so much that we didn't have as many flats to fix, but we still had some.

"After we moved down to that area with the Marines, several things were different – we ate with the Marines while we were there, for one. In the Seabees we were all just kind of alike, except the commissioned officers, but even non-commissioned officers in the Marine Corps -they don't eat with the privates. They don't associate with privates. After we moved in there with the Marines, they wanted us to eat in the separate chow hall too. The difference was, in that officer chow hall, those Japanese waitresses brought our food to us. When we ate there with the officers, we had to pass through the area where the Seamen and the privates were eating to get to the place where we ate."

Dad started smiling. "Me and Clifford would go through there with our noses turned up and they would boo and carry on. We didn't do it all the time, but we had fun with it.

"Another thing, they had a separate beer hall for the officers. They were even given a ration of whiskey. They could drink the hard stuff, but the privates, they just issued them beer. So once in a while me and Clifford would go to the 'exclusive beer and liquor hall.' But I never drank any; I didn't like the stuff. We'd go just to see what was going on."

"Sounds like you and Clifford had a good time," I said, as I set my glass down, then swept the crumbs off the table into my hand and emptied them into the trash can.

"We did. I know I've told you the story about us going out one night," he said, glancing at his watch and retrieving his cup from the microwave. "But I don't know if you have time for that."

"Of course I do." We both sat back down at the table. "I want to hear about everything else that happened and your trip home too."

"If you're sure."

I nodded and waited.

He crossed his hands, elbows on the arms of his captain's chair, a smile already playing around his mouth, and began again.

"When we were in Japan we could go out on liberty after a while, but we weren't supposed to be out at night. Me and my buddy Clifford knew one of the Shore Patrol or MPs, as the Marines call theirs. He was in our outfit, so one day we went down to the Shore Patrol office and hitched a ride with them into town - against regulations.

"Of course, the other MPs didn't know us and when they spied us out after dark, they took off after us. Clifford was a little taller than me, but we were both long-legged and could run. The part of town we were in was built against a slope, with the houses built right up against one another. I don't know what the Japanese inside those houses thought when they heard us runnin' across their tile roofs. We jumped from one house to another, and with our long legs, those MPs weren't about to catch us!"

Dad's eyes twinkled as he recounted his adventure.

"We ran until we finally came to a cemetery. The Japanese cremate, so their cemeteries are not like ours. The markers are real close together and some of them are head high. We went through that cemetery standing behind some of those markers. It was night and the MPs couldn't see us. We made it back to the barracks without getting caught. If we had, we would have gotten in trouble and we wouldn't have got to come home when we did."

I laughed at the mental image his story created and reminded myself that he was just a youngster of twenty-two at the time. I'm sure there are other escapades that I will never hear about, but that's as it should be. We all need our secrets.

Just then, Dad's face looked younger to me somehow...maybe it was the happy memory of his friend Clifford and youthful antics, or some inner source of joy that sparked that look...I didn't know, but I was glad to witness it all the same.

"You said you and Clifford enjoyed your time there in Japan...were there any special people besides the interpreter and his family that you got to know?"

He nodded and smiled.

"When I was in charge of the roads and maintenance of the truck tires, I picked some of the Japanese to help. I heard that the general population, the Japanese men, had to donate two days a month to the clean-up. Where I was, the Japanese would come every mornin' and row up, about four deep in a long row, and the ones that were in charge of details, like myself, would pick out who we wanted to use for the day. I needed someone to fix flats and there was one older man, probably in his forties, maybe fifties, that I started using. Every day he would wave his hand and jump up and down so I would see him and pick him. He worked for me until I left to come back to the States. When he found out I would be leaving soon, he brought me a present."

Dad made a circle with his thumb and fingers. "It was a little tea cup and saucer. When I told him I was going back to my country...he had tears in his eyes. He really took a liking to me and he was a good worker. He really took care of them tires."

Dad's face softened at the memory while he stared at the floor.

"Bet he did miss you…you probably treated him nicer than most did."

"Maybe. Some of the guys really resented the Japanese and hated them, but they were people who were just doing what they were told, like I said before."

"You want to switch channels as you say?" I quipped, and smiled at him, thinking he might want to lighten the mood.

He smiled back.

"Sounds good," Dad said. "I'll tell you about my one and only mail plane experience. I wanted to do some sight-seeing, so I hitched a ride with the mail plane over the island of Kyushu. The plane was a C-46 or 47, a two-motored cargo plane and it could carry a lot of stuff. On this particular run, it was overloaded. They didn't go over the mountains; they just weaved around through them.

"When we hit air pockets, as they call them, the plane would fall and I would come plumb up off the seat. When the plane finally caught, the wings would shudder like they were going to come apart. We made six landings on bombed-out air strips; it took all day to make the rounds. One guy got sick during the ride and messed up the airplane…the pilot was not happy." Dad grimaced. "By the time we got back, I decided that was enough for me. I wasn't going to tour any more on Kyushu!

"Another trip I made was down to Obama where there was some world famous hot springs. Me and some other guys took one of our 6-by trucks and went down there. It took several hours; the roads were terrible. The resort where we stayed was beautiful though and I spent three days there, enjoying the springs. Hot, boiling water ran out of the ground over some rocks into a glassed-in sittin' pool. You could ease into the water on the cooler end and gradually move towards the hotter water. Lookin' down at your body, you could see it turning red. When I got out an hour later, I could hardly stand up!"

"I bet that was a nice break, though, from your regular routine."

He nodded. "Yes, I enjoyed my stay there. I saw a lot of things I'd never seen before. The Japanese bedding is another interesting thing. Even in that nice resort, you slept on the floor. The mat I had, which was nothing more than bamboo and some batting, wasn't long enough. Even catty-corner I couldn't get all of my 6'2" frame on that mat, and the rice pillow would crawl out from under your head as you breathed. Several times I woke up and my head would be on the floor and I'd have to push that pillow back under my head. That bedding was definitely not made for tall Texans!"

"Bet you didn't get much sleep!" I exclaimed, then remembered something I'd wanted to ask. "Tell me how you and Paul Ludeman met. You and Mom were friends with him and his wife Lorna for many years. I remember looking forward to their visits; they always brought me and Josh something from Michigan."

"Paul was a replacement that came to Japan after the war was over to be a part of the Occupational Forces. He took a liking to me because I appointed him mail carrier. All he had to do was drive to Sasebo once a week to pick up our mail."

"No wonder he appreciated you!" I laughed. "I would too with a job like that."

"I didn't know Paul Marabito until we met at one of our reunion get togethers, but he was there at Omura too. We figured that out when he started describing the rock crusher there and a crematorium that was in the area. I knew right where it was, and of course, I was back and forth to the rock crusher while workin' there too. Funny we never met till nearly fifty years later.

"But Clifford and I were together all the way through, from Camp Endicott in Rhode Island to Japan. I didn't see him again until one of our reunions in Washington, D.C. - I believe it was in 1995. I walked into the hotel ballroom where our meeting was and Clifford spied me across the room, jumped up and came over to meet me. I recognized him in an instant. He looked the same, tall and lanky…just an older version of himself. We hadn't seen one another in almost fifty years, but we recognized each other immediately. I sure was glad to see him," Dad said wistfully.

"I know he was glad to see you too. You certainly went through a lot together."

He sighed. "Yes, we did. We're just about all gone now. There's just Paul Marabito and myself left, that I know of, but I haven't kept up with everyone, so I don't know for sure."

Wanting to shift onto more cheerful ground, I asked, "What about when you shipped out from Japan? How did you get back home with your sea bags and that big box Larry Brown made? You couldn't have carried all three the whole time?

"When we were fixing to ship out from Japan, I knew they were going to shake us down and take my G.I. stuff away from me. I had bought an extra one of those good fleece-lined Navy coats they had issued us and some other stuff and I didn't want them to take it. I put all my G.I. stuff in one sea bag, and all of my personal stuff in the wooden box and other sea bag. They lined us all up in rows, with our bags in front of us."

Dad leaned over and swiped his hand back and forth across the table top as he talked.

"There was a bunch of us leaving Sasebo - lots of rows - so I rolled my bag with the G.I. stuff in it into the row behind me till they passed by and inspected my bag, then I rolled it back in front of me. I guess that was foolish really…I wrestled with those bags all the way back to the States. It caused me to bend my Japanese saber that I brought back…trying to handle all that heavy stuff. I bent it when I jumped off a truck. We could also bring one Japanese rifle back, and ship one, so that added to my load. I guess I should have been like a lot of them and been happy-go-lucky and left all that stuff behind, but I grew up in the Depression and all the stuff I had was hard to come by." The Depression surely left its mark on Dad, he would save bits of barbed wire and rope, odd and end pieces of lumber, and straighten a nail to try and salvage it before he would get a new one. Without consciously trying to, Dad taught all of us kids to be savers too. Now, he's passing that frugal tendency to the next generation. Just the other day, I saw him showing my granddaughter how to roll up a piece of rope she had found while playing, I overheard him saying; "Now you save this because you never know when it might come in handy." She is learning from the best.

I finished off my water and checked my watch. It was getting late, but I didn't want to leave until I heard the rest of Dad's story.

"What about the trip back to the States from Japan? Tell me about that."

"I set sail for home along about the middle of March 1946 on one of our newer troop ships. We were somewhere between the Aleutian Islands and Hawaii when the seas got rough, and we found out from the ship's radio that there had been an earthquake in the Aleutians. The usually peaceful Pacific got worse and worse. The waves got higher and higher.

"Before they ordered us below deck, I got up on the main bow and took the ride of a lifetime. The weather was warm and the water was warm. I stood there ahold of the railing as the ship quartered into the waves. As one huge wave rolled under the front of the ship, the bow would rear up until it looked like about 150 to 200 feet down to the water. When that rolling wave reached the stern, the ship would fall into the valley between the waves and I could look up and see a wall of water above me. When the waves begun to completely cover the ship's deck, the captain feared we would be swept overboard, so he ordered everyone below.

"The ship was tossing so much that the cooks couldn't make hot meals because the contents of their big stainless steel pots would pour out. We went on dry rations. We were shut up in the belly of that ship for three days and three nights before the waves subsided. I thought about Jonah and how he must have felt shut up in the belly of that whale, but our accommodations **were** a little more comfortable.

"The ship was creaking and popping so much during those three long days and nights, that I thought, *After all we've been through, what a shame it would be if we were sunk by one of those huge waves and were all lost at sea*. There were about 3,000 of us on that ship. I'm glad it was a new one because it held together, even in those giant waves.

"While we were down below, we had to run our belts around part of our bunks and then our bodies to keep from being tossed to the floor. One of the guys didn't fasten himself in and got thrown out of his bunk and broke his arm.

"The piano was in the chow hall, fastened to the bulkhead with ropes, but the ship was heaving and tossing so much that the piano came up over the top of the ropes and broke into pieces. Several men were hurt by the flying debris."

"Boy, that's something. Go all through the war and then get wounded by a piano!" I exclaimed.

"Yeah, it was a wild ride alright. Later, we learned that the same tidal wave we rode out had rolled over Hilo, Hawaii and washed a lot of the town away."

"Wow…it's a miracle you made it home," I said shaking my head. "What else do you remember about your trip home?"

"Well, for one thing, even though it was an almost new ship, they tried to make us chip paint on the way back, but we didn't do much work," he said with a smile. "Where there were rust spots, we would chip the peeling paint off and prime it."

"Did you have any fun on the ride home?" I asked.

"I did have one fun job while I was aboard ship, and that was washing down the deck. I didn't mind that. I used a hose, kind of like a fire hose and that thing had power. 'Clean and sweep down all decks! Fore and aft!' they would holler over the ship's horn.

"One time I was on that detail, and the guys were scattered all over the deck I was supposed to be washing down. They were playing cards and everything." Dad got a twinkle in his eye. "I told them to *Clear the deck!*

They didn't and I turned that hose on them." He threw his head back and laughed. "They called me lots of different names."

"I bet they did!" I laughed along with him.

"One day I was on deck and I heard a familiar voice. I could pick it out because the fella it belonged to had a speech impediment. I walked over to where this bunch of guys were playing cards and I asked the fella I'd heard, *Are you Dud Moss?* He said, 'I surrr am!' Dud and I went to Enterprise community school together. I was surprised to see him. There we were on that big troop ship in the middle of the Pacific and I run into a boy from home. That was something."

Just then, Mom called out from the Big Room and I went to see what she wanted. Dad said he needed some more water before bed, so he got up to refill his glass from earlier in the day.

When I got back in the kitchen, I saw that he had broken off another piece of a fried pie. His slightly guilty expression made him look like a kid caught with his hand in the cookie jar.

"Hard to resist, aren't they?" I teased.

"Some things are not worth fighting," he deadpanned.

I laughed and reached for the other half of the pie. "I agree…now where were we?"

"What did your mama want?"

"Just wanted me to change a lightbulb; one of the ones in her ceiling fan had gone out again."

"Those things just don't last in that fan for some reason."

"I know…bet she's gone through a whole pack since March." I licked some crumbs off my fingers, sat down and asked again. "Now where were we?"

"In the middle of the Pacific, I think."

I nodded and waved my hand for him to continue. He leaned back and chewed thoughtfully for a bit before he answered.

"To pass the time on board ship, I read a book on finances - how to establish credit at a bank and how to manage money. It stood me in good stead when I got home. One valuable thing I learned was you don't wait until you need some money to borrow a little. Borrow some, pay it back on time, and then when you need it, you'll have some credit and know your banker. They need to teach kids how to manage money. It would help them a lot."

"Sure would…" I nodded. "Where did you dock when you got to the States?"

"San Diego. It was a good feeling to get back home. We were up on the deck, seeing what we could see when we came into port in San Diego. I remember thinking, if they can just get me to the West Coast, I can **walk** home from there. I was eager to get home. An officer asked me if I wanted to re-enlist - they would bump me up a notch, he said. But shoot, I was ready to be home.

"I made it all the way to San Diego with all my stuff, two sea bags and that box, but when we were getting ready to leave there and board the train, they shook us down again."

"Good grief! Looks like after all you poor guys had been through, they could let you keep a few things."

"Yeah, at the base, I saw the employees' kids runnin' around with those G.I. knives and stuff on their belts, yet they were taking them away from us!"

"Well, that wasn't right!" I fumed. "I don't blame you for wanting to keep your stuff. I can't believe the audacity of those parents."

"Well, somebody told me about a place just off the base where I could ship something C.O.D. So I went down there at 11:00 at night. They were still open and I shipped that G.I. bag home C.O.D.... no problem. The people shipping it were civilians. But I still had my G.I. issued knives in that wooden box, in the bottom under some stuff. And good thing I put it on the bottom because they shook us down again at Houston before we left the base. They opened it up, shoved stuff around a little bit, but they didn't find the knives. After they shut it up, I took off and managed to get home with my G.I. knife and another big knife I had acquired somewhere."

Dad kind of chuckled. "That beat all," he said, shaking his head.

"Where did you get the extra rifle and the saber you mentioned earlier?"

"The Japanese, as part of the surrender, had to turn in all their weapons. The military had stored them in a building near where we were staying, and I found out about it. Me and a cohort, I can't remember who, climbed up on the building and went through an upstairs window, and I "procured" another rifle and that saber."

"Glad you didn't get caught...you could have been in a lot of trouble, I bet," my eyes wide at the thought.

Dad agreed. "Young and a bit reckless, I suppose."

That story reminded me of something I wanted to ask about.

"You know that old mechanical pencil sharpener, the one on the wall over there by the fridge?" I said, gesturing in the direction of the refrigerator. "I know you brought that back from Japan too, but where did you get it?"

"We were cleaning out the American officers' quarters there in Japan, and that pencil sharpener and that cloth measuring tape I've used for years was in there. That was part of that G.I. stuff I had in my bag that I wanted to keep."

I smiled. "I'm glad you kept it; we've used that old pencil sharpener as long as I can remember." Over fifty years old, and it still works better than anything you can buy these days, I thought, before voicing my next question.

"What about the trip home on the train? What was it like?"

"The trip from San Diego to El Paso was about like the trip out there a couple of years earlier, but in El Paso they put us in cattle cars for the trip to Houston."

"What?! You guys went through the fire, and that's the best they could do when you got back?!"

Dad shrugged. "It was a terrible train, but they were using every set of rails we had - all the trains available because there were millions of G.I.s coming home. When I finally got to Houston, I got my discharge at Camp Wallace on April 10, 1946.

"An officer there asked me if I had any complaint against the service, and I told him they had messed with my teeth and I would like to get them fixed, so he signed me up to take care of them when I got home.

"When they turned us loose at Camp Wallace, we were just on our own. I was at the edge of Houston, walking across the railroad tracks, when I saw a little hamburger joint. It was one of those in a trailer, a little silver diner like they used to have, and the feller cooking in there was a boy I grew up with, Maxey Davis. He was the first person I saw that I knew. We went to school together at Enterprise. Maxey was a big fella; I couldn't believe there was ole Maxey flippin' burgers. He was too busy to talk much. I was hungry since I hadn't eaten for almost two days. I ordered a burger - it sure was good. Maxey wouldn't let me pay for it, said it was on the house."

"That was nice of him," I said, thinking that it still didn't make up for the cattle car.

As Dad answered, his left hand began that characteristic smoothing motion on the varnished table top, round and round as he talked.

"Yeah, I appreciated that."

"Where did you go from there and how did you get home?

His hand still going round and round, he continued. "About eight or ten of us decided that instead of taking a bus to get home, we would hire what they called a *traveling bureau*. It was a big over-sized van is what it amounted to. We all chipped in and got him to carry us to Tyler. Of course, they dropped me off in Jacksonville because it was on the way. He drove straight through, so we were in Jacksonville in about four hours."

"What did you do when you got to Jacksonville?"

"Well...I was lookin' the town over, standing on the street corner across from City Hall, on the west side, when Leo's sister Deliter Thorne and her husband Lawrence saw me, stopped and took me home. It had been about two years since I'd been home. I was overseas a year and a half. It was a wonderful feeling to be home...and to see Mama.

"But there was a big change after I settled in, a psychological change I guess. I had been *The Wheel* as they called me over in Japan, I had been in charge of a lot of things and had an important position - for a peasant boy - but when I came home, I wasn't anything except unemployed.

"I was still glad to be home, but I didn't have a purpose. That's why some men re-enlist, I suppose; they have a structured life in the military and for me, I felt my time in the service was really important because I was part of the structure in Japan.

"Having that kind of job made me feel pretty important; I would get everybody to work, and then I would just ride around and play big-shot. I guess I'd gotten kind of spoilt. I had my own private way to drive around and sightsee...and then, I got home and didn't even have a way to get to town. All the vehicles were worn out, and they hadn't made any for a big part of the war time - all the plants were busy making military vehicles. They started back making some automobiles after the Germans surrendered in 1945, I think, but a good car or truck was hard to come by.

"I guess the way I was feeling was the way a lot of the servicemen feel, even now, the struggle they go through when they come home. A little lost...frustrated and having to deal with what they did and saw in battle; it can work on a fella's mind.

"I'm glad I finally went to G.I. school and started workin'; that helped me adjust, and eventually I was able to put the war behind me and live my life. I had nightmares for some twenty-five years, though, before they stopped. War changes you, but I hope I've used my experience to make my life better, at least I'd like to think so."

Dad sat quietly with his head slightly bowed as if lost in thought. I just sat and looked at him for a minute, trying to name what I was feeling. Part pride, part wonder…some sadness for what he went through, but mostly I just felt love. A fierce overwhelming love for this amazing man I was blessed to call Dad. With all he had been through and the horrors he had witnessed, he was still the gentlest, kindest, and wisest man I'd ever known.

I got up out of my chair, went over and without a word, gave him a hug. He patted me on the shoulder. "You are a worthy daughter," he said. Those were the most beautiful words I've ever heard.

Ever since I stepped off the train in Houston, I've had the nagging feeling that coming home wouldn't be what I expected. But after seeing the familiar streets of Jacksonville and enjoying the ride out to our road, I am more settled.

When Lawrence and Deliter dropped me off at the end of the lane, I felt a sense of relief start to come over me and the closer I get to the house, the more relaxed I feel.

So many memories of growing up flood my mind as I walk this red dirt road. I don't ever want to leave East Texas again.

As I round the last clump of elms and blackberry bushes along the ditch by the home place, the first thing I see is the pecan tree that Grandma Tab planted. Looking up into the still bare branches, I am glad I will be here to see it leaf out. It has grown since I've been gone. Bending down and searching the trunk, I finally find the horizontal scars about two feet up and run my fingers over them. Smiling, I remember using my dad's trapping axe to try and cut it down. Thank goodness Mama stopped me and put an end to my five-year-old logging ambitions.

Just beyond the pecan tree I see the old box-plank house. The weathered bare boards streaked with gray, the half-and-half roof dad had made. One part corrugated tin that came off that old house he bought in Tecula when the cotton gin closed. The other half wood shingles dad had made over at Cribb's mill before the war.

It is the most beautiful building in the world to me right now.

As I walk the dozen feet or so to the gate, I notice a couple of quilts are on the yard fence to air in the sunshine, and I see that Eva's rose bushes alongside the front gate have some pink buds on them.

Reaching for the gate latch, I hear the slap of the screen door and look up as I step through the gate.

Mama is carrying another quilt to the fence.

Shooing the hens out from under her feet, she walks over and shakes out the quilt before carefully draping it over the fence. As she arranges the quilt, the spring breeze teases a strand of her silver hair from the familiar braid atop her head. When she reaches up to smooth it back, she catches sight of me.

Putting her hand up to shield her eyes from the sun, I see her other hand fly to her heart...she knows I'm home!

I set my sea bag down inside the gate and start walking toward her. I'm tempted to run, but I want this moment to last as long as it can.

When we get close enough to touch, we both stop, searching one another's eyes, looking for that bridge that will erase all the months of loneliness and worry.

I see it first...in the way she smiles. It starts in her eyes and flashes across her face in an instant. She holds out her arms and, like the boy I used to be, I fall into them wholeheartedly.

I am home...really home...the place where I feel safe and loved and where someone will always greet me with open arms. It is heaven.

GLOSSARY OF TERMS

Traveling bureau — a van-like vehicle used to transport people during the 1930s and '40s

Seabee — a common reference to persons in the Naval Construction Battalions during World War II

CB — abbreviation for Construction Battalion and combined with a number, it refers to a particular unit (Dad said the Marines liked to joke that CB stood for Confused Bastards.)

Big Thicket — a densely forested area in East Texas now a National Preserve.

CCC — abbreviation for the Civilian Conservation Corps created in the 1930s by President Roosevelt

KP — military term meaning kitchen patrol or kitchen police

GI —military term meaning government issued or general issue

Quonset hut — A large building made of curved corrugated metal panels created for use during World War II. (The Seabee Museum in Rhode Island contains seven examples of vintage Quonsets that were rescued from demolition. The museum stands on the site of the WWII Camp Endicott.)

Head — Navy/Marine slang term for bathroom

Carbine — term used for guns with a barrel of less than 18 inches; rifle refers to guns with a barrel of 18 inches or more. (Carbines were easier for the Seabees to carry, since they had to work with the guns on their backs, at the ready.)

LCI — Landing Craft Infantry is a term used for ships that were used to carry troops. They were big enough to cross the Pacific, but small enough to land a company of infantry on the beach. Used all over the Pacific during WW II, the LCI was usually outfitted with five 20mm cannons; some also carried rocket launchers, mortars, or .50 caliber machine guns.

WAC — stood for Women's Army Corps, a branch of the Army created during WW II to enable women to serve in non-combat positions

LST — Landing Ship Tank was the workhorse of the US Navy during WW II. They only had a number designation instead of a name and were large enough to carry twenty Sherman tanks on deck

Higgins boat or **LCVP** (Landing Craft/ Vehicle/ Personnel) — an amphibious vehicle, the invention of Andrew Higgins, and is credited with helping win the war. Used on D-Day, Iwo Jima, and other battles, the Higgins boat proved invaluable. The Higgins boat could hold up to 36 troops and could carry up to 8,100 pounds of cargo.

Tokyo Rose — the name given to English-speaking Japanese women who broadcasted over the radio during the war in the Pacific. Credited with spouting Japanese propaganda, the infamous Tokyo Rose broadcasts were listened to by sailors and soldiers alike. Once thought to be attributed to one woman, after the war, it was discovered that there were several women who were on the air during that time.

APA — Amphibious/Attack Transport Ships designed to carry troops and support equipment. It could hold an entire battalion of troops (anywhere from 800 to 1,000 men).

6-by — Originally named 6x6s, these vehicles were all-purpose trucks used to haul cargo or people; equipped with a divider in the bed that could be raised and locked into place, converting it into a dump truck to haul rock, gravel, dirt or clay. The truck had double axles in the rear with four wheels on each side and two tires in the front with a total of ten tires on the ground. Its load capacity was six tons. The servicemen referred to the 6x6's in the shortened version: 6-by.

Parade Article

This is a reprint of an article originally posted on *Parade.com* for Father's Day 2017. I wanted to include it in this book to give the reader a glimpse into Dad's personality and why he means so much to our family.

> My Dad is a happy man. It wasn't until I was grown that I began to realize how remarkable that was. This summer Dad will turn 94. He grew up during the Great Depression and was raised on the values of "Wear it out, use it up, make do or do without;" a philosophy forgotten in today's throwaway society. He served his country during World War II, fighting in the Battle of Iwo Jima. Dad witnessed the famous flag-raising on Mt. Suribaci.* He survived two months in a foxhole in horrible conditions and witnessed indescribable carnage. Yet, I have never seen anyone as positive and full of appreciation for life as my dad.
>
> Father's Day has always been extra special for me because it was a celebration I shared with him. My birthday always fell on or close to Father's Day, so we just combined them into one family event.
>
> I was the middle child of three girls and two boys, but I never felt overlooked or slighted in any way. Sharing was just something we all did.

Dad worked hard as a painting contractor for over 50 years, but I never once heard him complain, even when he worked with a bout of bleeding ulcers, and later had to have surgery. He believed in working hard, but he always took time to notice and appreciate the everyday things in life. We all looked forward to him coming home from work and listened for the sound of his truck pulling into the driveway. All five of us kids would stampede to see who could make it to the back door first to meet him. I can still see him, his 6-foot-two frame filling the doorway, black eyes twinkling, head thrown back in laughter at our exuberant greeting.

His sense of humor is keen, and he has a way of phrasing things that made them memorable. Dad enjoys having fun, but never at anyone's expense. He taught us compassion and respect by the way he treated others. Growing up, we had many people pass through our rural area, drifters, "fellas down on their luck" Dad called them. I watched him give them odd jobs around the farm, he would explain what he needed done, clearing fence rows, piling brush or chopping firewood, then, he would always shake hands on the deal, looking them square in the eye. I don't recall a single one that left without finishing the job.

We weren't wealthy monetarily, but listening to Dad, we felt rich. On grocery day, usually Fridays, Mama would bring home 8 or 9 big paper Piggly Wiggly sacks full of flour, Crisco, crackers, sugar and maybe a treat or two for us. My brothers, sisters, and I would help Mama tote them into the kitchen and soon, our green linoleum floor would be over-flowing with scattered sacks and groceries. Dad would sometimes come home in the midst of the chaos and stand at the door, hands on hips grinning and say, "Well! Where did all this come from? Rich folks must live here!" And we did. We were rich in a deep down, secure loved way that is still with us, all these years later.

Dad rarely verbalized his life lessons; they were more 'caught than taught'. He is a stickler for honesty and guarding his reputation. In fact, in our small town, Dad's reputation is so stellar that the bank president gave Dad the keys to the bank and the vault, so Dad could go in over the weekend and paint without interrupting the regular business of the bank.

We've had some tough times over the years, losing one of my brothers; many extended family members, and financial hardships, but the closeness and loyalty we feel for one another has never wavered. He has shown us what true love is, by caring and loving our mother for over 70 years now.

None of us realized it growing up, but while we were helping Dad with the garden, the cows, hauling hay and building fences, barns, and corrals, he was showing us how to live a good life–be a real man and father, and a person of character.

Dad loves Thanksgiving and Christmas. He manages to make the holidays magical through the traditions he's kept through the years. He always cuts some holly and hangs it over the doorway in the kitchen signaling the beginning of the Christmas celebration. When we were kids, he planned the Christmas tree hunt. Every year, he would drive to a place in our woods where he had spied some likely candidates and then we'd spend several hours tromping around, comparing the virtues of different trees until we decided, by vote, the winner. Never in a hurry, Dad would stroll along with his axe on his shoulder, relishing the adventure.

Dad and I also share a love of coffee, horses and travel. He likes to say "I enjoy seeing what's over the next hill." I give him credit for my "itchy feet." It has always amazed Dad that he could 'travel the world on a poor man's salary.' In fact, it was only in his later years that he and Mom could afford to visit Europe and Israel, a life-long dream of theirs. Raising all of us wasn't cheap, with braces, college tuition and weddings, but as Dad says, "It was all worth it, you kids are my life."

We also loved to ride our horses down the quiet country lanes near our farm. I enjoyed having Daddy all to myself for a little while. Secretly, I was glad that my brothers and sisters didn't care for horses. Later, when I started riding competitively, I could always spy Dad in the stands, cheering me on. He also put up with me bringing all kinds of animals home to nurse back to health, both domestic and wild. He once nicknamed me "Dr. Doolittle" because I had such a rapport with animals.

Dad and I were the early risers in our house. My earliest memories are of us sitting out on our screened-in porch watching the sun come up, sipping our coffee. Mine was mostly cream then, but I didn't know or care. I felt so big and important sharing that time with Dad. He has a way of making all of us feel special.

Yesterday, I went over to have my afternoon coffee with Dad, which has been our routine for many years. He was already sitting in his chair on the patio out back. It's the perfect spot for admiring the beauty of the rolling hills and ponds on the farm. I grabbed a cup from the kitchen, and then joined him, settling into the chair beside him. His eyes never left the peaceful scene before him. "You know," he said, "I've seen many a beautiful sight in my lifetime, but this is one my eyes never tire of." As I turned to study his profile, his shock of white hair and weathered face, coffee cup poised in mid-air like he was posing for an artist's sitting, I took a mental snapshot. Soaking in that portrait of my dad in that moment, I tucked it away in my memory bank. Smiling, I said, "Dad…I couldn't agree with you more."

*The first flag raising was the most important to the guys on the island because it meant they had taken Suribaci, even though the second flag raising, photographed by Rosenthal, was the one that captured the nation's imagination and became a symbol of the war in the Pacific.

About the Author

Tamra McAnally Bolton is a photojournalist who has written for numerous publications both in print and online. Her photos have appeared in dozens of magazines and she has won awards for both her photography and writing. She has spent almost two decades collecting stories of WW II veterans and working to preserve their personal experiences. Tamra is passionate about travel, history, and being a story-keeper for her family.

Contact

You can contact Tamra through the publisher at:
povertyridgepress47@gmail.com

Made in the USA
Coppell, TX
01 February 2020

15294840R00122